Information Circular 9529

Coal Dust Explosibility Meter Evaluation and Recommendations for Application

Marcia L. Harris, Michael J. Sapko, Floyd D. Varley, and Eric S. Weiss

DEPARTMENT OF HEALTH AND HUMAN SERVICES
Centers for Disease Control and Prevention
National Institute for Occupational Safety and Health
Office of Mine Safety and Health Research
Pittsburgh, PA • Spokane, WA

August 2012

This document is in the public domain and may be freely copied or reprinted.

Disclaimer

Mention of any company or product does not constitute endorsement by the National Institute for Occupational Safety and Health (NIOSH). In addition, citations to Web sites external to NIOSH do not constitute NIOSH endorsement of the sponsoring organizations or their programs or products. Furthermore, NIOSH is not responsible for the content of these Web sites. All Web addresses referenced in this document were accessible as of the publication date.

Ordering Information

To receive documents or other information about occupational safety and health topics, contact NIOSH at

>   Telephone: **1–800–CDC–INFO** (1–800–232–4636)
>   TTY: 1–888–232–6348
>   e-mail: cdcinfo@cdc.gov
>
>   or visit the NIOSH Web site at **www.cdc.gov/niosh**.

For a monthly update on news at NIOSH, subscribe to NIOSH *eNews* by visiting **www.cdc.gov/niosh/eNews**.

DHHS (NIOSH) Publication No. 2012–172

August 2012

SAFER • HEALTHIER • PEOPLE™

Table of Contents

Executive Summary 1
Introduction 3
Background on Coal Dust and Explosibility Testing 4
CDEM Operation 6
Comparison of Laboratory Results and CDEM Results 7
Joint Study between NIOSH and MSHA 8
Results and Discussion 9
      GREEN Measurements 13
      RED/YELLOW Measurements 16
Conclusions from the NIOSH Study 20
Commercial CDEM Development 21
      Calibration and Programming of the Commercial CDEM 22
      Commercial CDEM Changes based on Potential Customer Concerns 22
      The Commercial CDEM as a Verification and Compliance Tool 23
NIOSH Recommendations 23
Acknowledgments 24
References 25
APPENDIX A: CDEM Design 27
APPENDIX B: CDEM Training 31
APPENDIX C: Prototype CDEM Calibration and Testing Procedures Used in the Joint Study 33
      Preparation of Calibration Sample 34
      Drying the sample 35
      CDEM Calibration 36
      In-Mine Testing 37
APPENDIX D: Particle Size Effect 41
APPENDIX E: MSHA Inspector Questions and Comments 45

Figures

Figure 1. Effect of particle size of Pittsburgh seam bituminous coal dust on the explosion propagation for % TIC as tested within LLEM. The dashed curve represents the propagation/nonpropagation boundary. 5

Figure 2. Coal Dust Explosibility Meter (CDEM). 6

Figure 3. CDEM measurement comparison of NIOSH-received samples. 12

Figure 4. Comparison of CDEM results and lab TIC results for one representative mine. 13

Figure 5. CDEM GREEN (G) measurement comparison of NIOSH-received samples. Agreeing GREEN samples have LTA results > 80% IC. Disagreeing GREEN samples have < 80% IC LTA results. 14

Figure 6. GREEN sample of 73.9% IC. 15

Figure 7. Comparison of MSHA and NIOSH incombustible measurements 17

Figure 8. CDEM RED (R) measurement comparison of NIOSH-received samples. Agreeing RED samples have LTA results < 81.5% IC. Disagreeing RED have more than 81.5% IC LTA results. 19

Figure 9. Agreement between CDEM and laboratory analysis. 20

Figure A1. Coal Dust Explosibility Meter (CDEM). 28

Figure A2. Extinction limit for normalized reflectance 30

Figure C1. Sample tube containing molecular sieves. 36

Figure C2. Collecting a band sample. 37

Figure C3. Sifting a sample. 38

Figure C4. Collecting a sample into a sample tube. 39

Figure C5. Transferring dust to a sample cup. 39

Figure C6. Testing a dust sample with the CDEM. 40

Figure D1. A comparison of coal dust particle size distributions on ϕ values. Rock dust particle size distributions are constant. 42

Figure D2. A comparison of rock dust particle size distributions on ϕ values. Coal dust particle size distributions are constant. 43

Figure E1. Cross section of a very thin (0.01-in-thick) explosible float coal dust layer deposited on top of a 3/4-in-thick (20-mm-thick) layer of rock dust. 51

Tables

Table 1. Elapsed time between dates that samples were collected and the MSHA inspector was notified of laboratory results 9

Table 2. Comparison of CDEM results as determined by separate MSHA and NIOSH testing 11

Table 3. Particle size distribution of 72.7% IC dust sample and LTA results for three different size fractions of the sample 16

Table B1. Mixtures of RD and PPC for CDEM calibration 32

Table C1. Pittsburgh pulverized coal size data 34

Table C2. Reference mixtures, percentage by weight 35

Acronyms and Abbreviations

CD	coal dust
CDEM	Coal Dust Explosibility Meter
CFR	Code of Federal Regulations
ETS	emergency temporary standard
IC	incombustible content
LLEM	Lake Lynn Experimental Mine
LTA	low temperature ashing
MSHA	Mine Safety and Health Administration
NIOSH	National Institute for Occupational Safety and Health
Φ	normalized reflectance
Φ^*	normalized reflectance at the extinction limit
OMSHR	Office of Mine Safety and Health Research
PPC	Pittsburgh pulverized coal
K	product of rock dust to coal dust particle density ratio and ratio of mean particle diameters of coal to rock dust
RI	Report of Investigations
RD	rock dust
SUT	sample under test
S_c	surface area of coal dust
S_r	surface area of rock dust
TIC	total incombustible content
USBM	United States Bureau of Mines
H_2O	water/moisture

Unit of Measure Abbreviations

ft	feet
f_r	fraction rock dust
g	grams
I_x	light intensity of coal and rock dust mixture
I_c	light intensity of coal dust
I_r	light intensity of rock dust
μm	micrometers or microns
mL	milliliters
mm	millimeters
oz	ounces
% IC	percentage of incombustible content
% CH_4	percentage of methane
% RD	percentage of rock dust
% TIC	percentage of total incombustible content
wt	weight

Coal Dust Explosibility Meter Evaluation and Recommendations for Application

Marcia L. Harris,[1] Michael J. Sapko,[2] Floyd D. Varley,[3] and Eric S. Weiss[4]

Office of Mine Safety and Health Research
National Institute for Occupational Safety and Health

Executive Summary

This report details the results of a NIOSH investigation on the ability of the Coal Dust Explosibility Meter (CDEM) to accurately predict the explosibility of samples of coal and rock dust mixtures collected from underground coal mines in the U.S.[5] The CDEM, which gives instantaneous results in real time, represents a new way for miners and operators to assess the relative hazard of dust accumulations in their mines and the effectiveness of their rock dusting practices. The CDEM was developed by the National Institute for Occupational Safety and Health (NIOSH) and successfully underwent national and international peer review. The intention of the device is to assist mine operators in complying with the Mine Safety and Health Administration (MSHA) final rule 30 CFR* 75.403, requiring that the incombustible content of combined coal dust, rock dust, and other dust be at least 80% in underground areas of bituminous coal mines.

As a final step towards commercialization of the CDEM, and to evaluate the performance of the device as a potential compliance tool, NIOSH undertook an extensive cooperative study with MSHA. This study, completed in 2009–2010, involved field use of the CDEM within MSHA's 10 bituminous coal districts. As part of their routine dust compliance surveys in these districts, MSHA inspectors collected sample coal and rock dust mixtures, field testing these samples for explosibility with the CDEM. Samples were then sent to the MSHA National Air and Dust Laboratory at Mt. Hope, WV, for parallel testing, first using a drying oven to determine the

---

[1] Research Chemical Engineer, Fires and Explosions Branch, Office of Mine Safety and Health Research (OMSHR), NIOSH, Pittsburgh, PA.
[2] Senior Research Physical Scientist (retired), OMSHR, NIOSH, Pittsburgh, PA.
[3] Chief (retired), Fires and Explosions Branch, OMSHR, NIOSH, Spokane, WA.
[4] Senior Research Mining Engineer, Fires and Explosions Branch, OMSHR, NIOSH, Pittsburgh, PA.
[5] The testing was conducted with a prototype version of the CDEM that was available in 2009. The commercial version, CDEM-1000, was released in 2011 and includes improvements in both software and hardware. These improvements are listed in the "Commercial CDEM Development" section and take into consideration many of the MSHA inspector comments detailed in Appendix E.

*Code of Federal Regulations. See CFR in References.

moisture followed by the traditional low temperature ashing (LTA) method. The LTA method determines explosibility of a coal and rock dust sample in a laboratory by heating the mixture to burn off the combustible material. The results, when combined with the moisture, are reported as total incombustible content (TIC). If the TIC is ≥ 80%, the sample is deemed to be nonexplosible and compliant with 30 CFR 75.403.

In the field component of this study, MSHA's use of the CDEM indicated that 30% (175) of the 591 samples collected were explosible. NIOSH was able to obtain and remeasure 297 samples, and 97% of those identified by the CDEM as being explosible (27% of samples) or nonexplosible (73% of samples) correlated with the results of the subsequent lab analysis using the LTA method. Of the remaining 3% where there were differences between the field and laboratory methods, subsequent NIOSH evaluation attributed these differences to the variability (incomplete mixing, inadequate drying of the sample, the particle size of the rock dust and/or coal dust) of the samples being analyzed, the retained moisture in those samples, and the inherent ash in the coal.

In considering these results and comparing the CDEM field measurements to the LTA laboratory measurements, it is important to understand the fundamental distinctions between the two methods. The determination of TIC by the LTA method is not itself a direct measure of explosibility, but a surrogate that calculates a single parameter associated with full-scale experimental results. This method is not based on particle size and treats all particles equally regardless of the size. In contrast, the CDEM utilizes a different approach, using optical reflectance to determine the ratio of rock dust to coal dust in a mixture, with full-scale experiments on flame propagation having already demonstrated the effects of varying the coal dust particle sizes and incombustible concentrations on the explosible vs. nonexplosible dust mixtures. A final important distinction between the two methods is that the CDEM offers real-time measurements of the explosion propagation hazard within a coal mine entry, allowing for immediate identification and mitigation of the problem, while the results from the traditional LTA method are not known for days or weeks after a sample is collected, allowing for the deficiency in rock dust to continue.

The conclusions of this study strongly support the field use of the CDEM to measure the explosibility of coal and rock dust mixtures, to more effectively improve the onsite adequacy of rock dusting for explosion prevention. Mine operators could use the CDEM on a regular basis to ensure that their rock dusting practices are achieving inertization requirements and meeting the intent of 30 CFR 75.403. MSHA inspectors could use the CDEM as a tool to immediately identify onsite explosibility hazards and initiate corrective action. A critical issue to both the LTA and the CDEM analysis methods is that the results are dependent on representative samples being collected for analysis.

Introduction

Federal regulations require that rock dust (RD) be applied in all underground areas of a coal mine to mitigate the propagation of a coal dust (CD) explosion. Prior to September 2010, U.S. Federal law 30 CFR 75.403 mandated that the nation's coal mines maintain a total incombustible content (TIC) of at least 65% in nonreturn entries and at least 80% in the return airways.[6,7] The 65% TIC requirement was based on an average particle size termed "mine-size dust," which was based on an average of representative samples collected from mines in the 1920s. To determine compliance with the federal regulation, mine inspectors systematically collect dust samples from sections of underground coal mines and send the samples to the Mine Safety and Health Administration (MSHA) National Air and Dust Laboratory at Mt. Hope, WV, for analysis of incombustible content.

The TIC analysis is a gravimetric (mass) measurement of the incombustible content (IC) of a coal and rock dust mixture. Generally, the analysis is attained using a low temperature ashing (LTA) method [NIOSH 2010]. Due to the inherent time needed to collect the samples, ship the samples, and then test the samples, the Coal Dust Explosibility Meter (CDEM) was developed to allow for immediate determination of the explosible reactivity of a coal and rock dust mixture. The device was tested using experimental coal and rock dust mixtures and on band samples collected by MSHA inspectors from underground coal mines [Harris et al. 2008].

In 2009–10, the National Institute for Occupational Safety and Health (NIOSH) and MSHA conducted an extensive cooperative study to contrast explosibility assessment as determined by the CDEM with explosibility assessment results as determined by the laboratory gravimetric analysis of incombustible content. Further, the study was able to evaluate the feasibility for inspectors to use the CDEM within 10 of MSHA's bituminous coal districts. This report will discuss the study results, with emphasis on comparisons of the CDEM explosibility assessment with the traditional method for determining the TIC. CDEM operation and the use and application of the commercial CDEM will also be discussed. Importantly, the study results are presented in the context of the current standard requiring not less than 80% TIC in all areas of an underground coal mine.

---

[6] In September 2010, MSHA published an emergency temporary standard (ETS) increasing the total incombustible requirement in intake airways to 80%. The final rule 2011-15247, requiring 80% TIC in intakes, was effective June 21, 2011. The current study was conducted prior to the ETS and subsequent final rule.

[7] Total incombustible content (TIC) includes measurements of the as-received moisture in the samples, the ash in the coal, and the rock dust. Incombustible content (IC) includes measurements of the ash in the coal and the rock dust and does not include the moisture.

Background on Coal Dust and Explosibility Testing

Coal dust particle size has a significant impact on the explosion propagation potential of coal and rock dust mixtures. As the coal dust particle size decreases, the coal particles are more reactive and increased amounts of compliant rock dust are necessary to render the coal dust inert.[8] Coal particle size has a high variability both within and between mines, with size being dependent on factors such as mine type (i.e., longwall or continuous miner), bit cutting speed, cut depth, and coal type. Size distribution will also vary along mine entries as coarser dust is deposited from ventilation streams closer to the production area, while finer dust is carried further down the entries.

Numerous coal dust explosion tests have been conducted in the NIOSH Lake Lynn Experimental Mine (LLEM) to specifically quantify the concentration of rock dust required to prevent flame propagation [NIOSH 2010]. These tests studied flame propagation as a function of coal dust particle size while using a rock dust particle size of ~75% < 200 mesh (volume median rock dust particle diameter of ~ 25 microns, or μm) (Figure 1). Based on these results, the greatest impact on explosibility is evident between the particle size of the 20% < 200 mesh coal (mean coal particle diameter of 96 microns, μm) and 80% < 200 mesh (mean coal particle diameter of 33 μm). To ensure nonpropagation within the LLEM, the 20% < 200 mesh coal dust required a 70% TIC (~ 68% rock dust) and the 80% < 200 mesh coal required ~ 81.5% TIC (80% rock dust) to prevent sustained flame propagation. Once the 80% < 200 mesh benchmark had been reached, no additional incombustible content (IC) was required to prevent flame propagation with further decreases in coal dust particle size under these full-scale experimental conditions.

---

[8] Compliant rock dust is defined in 30 CFR 75.2 as "Pulverized limestone, dolomite, gypsum, anhydrite, shale, adobe, or other inert material, preferably light colored, 100 percent of which will pass through a sieve having 20 meshes per linear inch and 70 percent or more of which will pass through a sieve having 200 meshes per linear inch; the particles of which when wetted and dried will not cohere to form a cake which will not be dispersed into separate particles by a light blast of air; and which does not contain more than 5 percent combustible matter or more than a total of 4 percent free and combined silica (SiO_2), or, where the Secretary finds that such silica concentrations are not available, which does not contain more than 5 percent of free and combined silica."

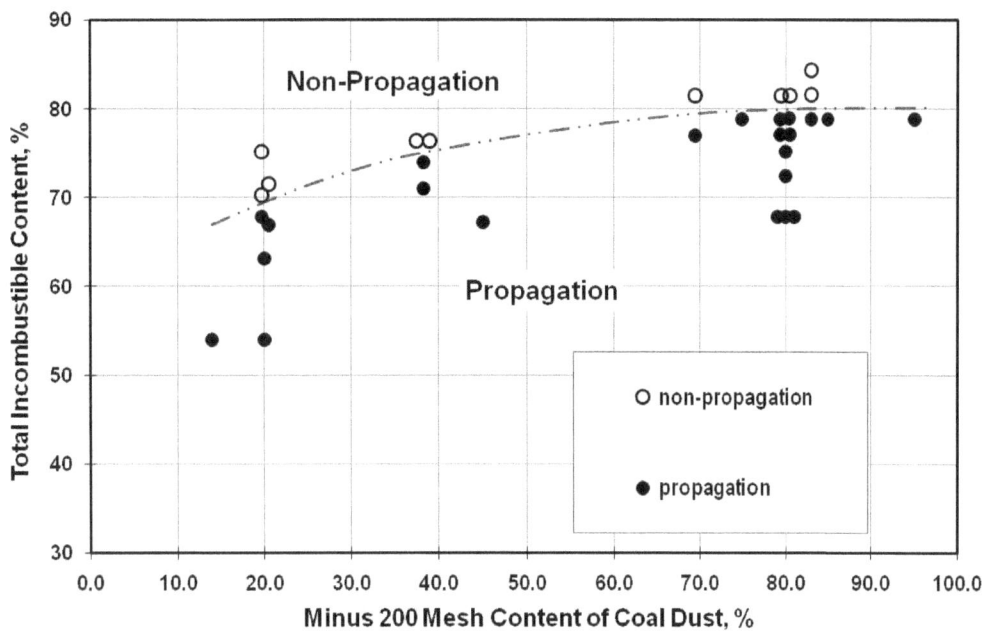

Figure 1. Effect of particle size of Pittsburgh seam bituminous coal dust on the explosion propagation for % TIC as tested within LLEM [NIOSH 2010]. The dashed curve represents the propagation/nonpropagation boundary.

To determine compliance with current regulations set forth in 30 CFR 75.403, inspectors from MSHA periodically collect samples of deposited dust from specified areas in a mine. The MSHA National Air and Dust Laboratory determines TIC and compares this TIC with the standard of 80% minimum TIC.[9] The TIC includes measurements of the moisture, the ash in the coal, and the rock dust. If 10% of the samples collected in a survey are < 80% TIC (in the absence of methane[10]), MSHA considers the sample survey to be noncompliant. The mine operator is issued a citation and a timeframe is stipulated to abate the citation. Abatement is accomplished by applying additional rock dust to the deficient areas from which the sample was collected. Abatement is confirmed through visual assessment by the MSHA inspector, but no followup dust samples of the abated area are collected or analyzed to ensure compliance with the respective intake or return airway TIC requirements.

---

[9] The 80% TIC requirement is based on explosion temperature thermodynamic limit models for coal and rock dust mixtures, extensive in-mine coal dust particle size surveys, and multiple explosion experiments at the Lake Lynn Experimental Mine [NIOSH 2010]. Presently, the size of the coal dust particles is not determined by the MSHA National Air and Dust Laboratory as part of the explosibility assessment.

[10] Per 30 CFR 75.403, "Where methane is present in any ventilating current, the percent of incombustible content of such combined dust shall be increased 0.4 percent for each 0.1 percent of methane."

CDEM Operation

The CDEM (Figure 2) is a handheld device developed to assess the explosibility of coal and rock dust mixtures in real time. The principle of operation of the device is based on the measurement of near-infrared radiation reflected from the surface of a homogeneous mixture of two dusts with different optical reflectance, in this case light-colored rock dust and dark coal dust. Near-infrared radiation is emitted by a light-emitting diode located behind the window of the CDEM probe.

When the CDEM probe is inserted in the dust mixture, the near-infrared radiation reflects off the surface of the dust and back to a silicon photodiode sensor. The normalized reflectance, Φ, is related to the rock dust to coal dust particle density ratio and the ratio of the mean particle diameters of coal to rock dust contained in the mixture. The normalized reflectance for the tested sample is compared to that of the calibration sample. If the test sample normalized reflectance is greater than that of the calibration sample, it is determined to be nonexplosible. If it is less than the calibration sample normalized reflectance, it is classified as explosible. For further detail on the CDEM design, calibration, and operation, see Appendices A, B, and C.

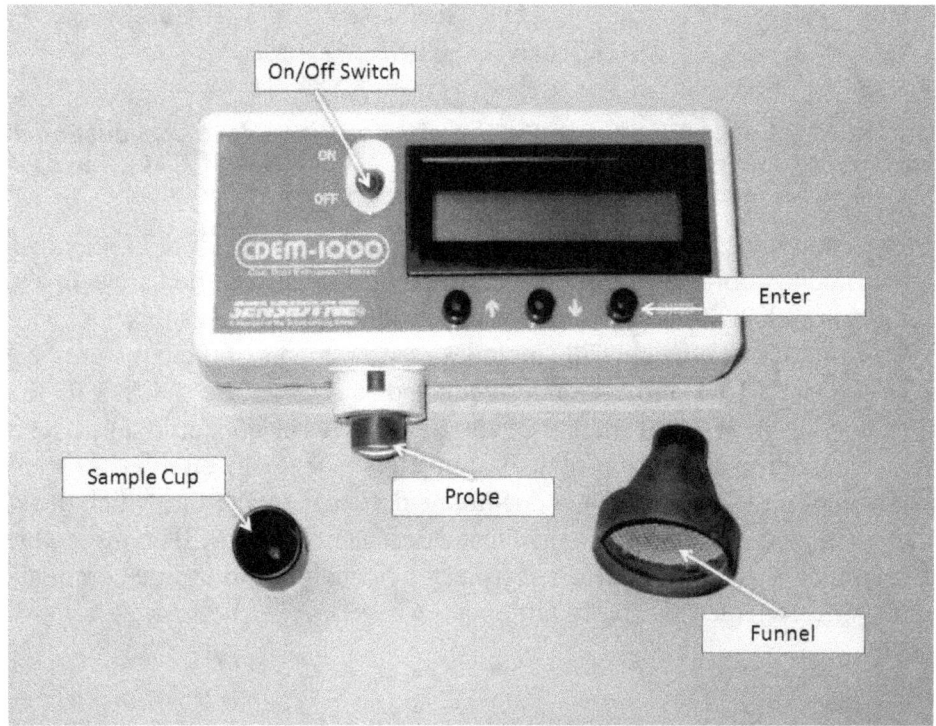

Figure 2. Coal Dust Explosibility Meter (CDEM).

Operationally, the CDEM uses a digital readout that identifies a sample as being either "RED" or "GREEN." For the prototype CDEM used in this study, the meter identifies a sample as RED (potentially explosible and requiring more rock dust) when the measured Φ is ~ 6% or more below the extinction limit Φ^* needed to prevent flame propagation (Figure 1 and Figure A2).[11] The extinction limit Φ^* is the boundary between propagation and nonpropagation and is set at the 80% rock dust level during CDEM calibration. The CDEM identifies a sample as GREEN (nonexplosible) when the measured Φ is equal to or greater than the extinction limit Φ^*.[12]

Comparison of Laboratory Results and CDEM Results

The traditional low temperature ashing approach to determine if a coal and rock dust mixture passing through a 20 mesh sieve (< 850 μm) is compliant with the inert requirement is significantly different from that approach used by the CDEM for assessing the potential explosibility of the coal and rock dust mixture. The current LTA method actually consumes the coal dust and considers the remaining material to be inert. Compliance with the law is then determined by comparing the measured percentage of inert material of the representative band sample with the pre-established requirement of 80%. The TIC of the sample includes the rock dust, the amount of moisture as received at the lab, and the inherent ash in the coal. The LTA method is not itself a direct measure of explosibility but is a surrogate that calculates a single parameter associated with full-scale experimental results and is also insensitive to particle size. In contrast, the CDEM determines the potential reactivity of the coal and rock dust mixture by optically comparing the ratio of the surface area of the rock dust particles to the surface area of the coal dust particles, and relates the measured ratio to a stored 80% rock dust/coal dust calibration sample at the extinction limit of Φ^*. The 80% calibration sample is prepared with the particular rock dust used at the coal mine and mixed with the standard Pittsburgh pulverized coal (PPC) dust [80% < 200 mesh (~ 74 μm)].[13]

When comparing methods that make accuracy determinations, the tendency is to compare the new method with the traditional method head-to-head. In this case, since the LTA method and the CDEM use different means to determine the explosibility of a coal and rock dust mixture, it is difficult to directly compare one method with the other. The only effective approach should be

---

[11] The commercialized version of the CDEM identifies a sample as RED when the measured Φ is below the extinction limit Φ^* needed to prevent flame propagation—i.e. when less than 80% rock dust.

[12] In addition to RED and GREEN, the prototype CDEM also identified a sample as YELLOW when the measured Φ was within 5% of the extinction limit Φ^*. A YELLOW reading indicated that the sample was marginally explosible. This feature was eliminated as unnecessary in the later-developed commercial version of the CDEM. However, in this report, the YELLOW readings were considered to be RED in the analysis of the data discussed in this report. Although the YELLOW measurements are not discussed separately, all of the RED, YELLOW, and GREEN measurements are included in this report.

[13] Pittsburgh pulverized coal (PPC) has been used in large-scale explosion tests at the Lake Lynn Experimental Mine (LLEM) and is the standard upon which the 80% total incombustible content cited in 30 CFR 75.403 is based [NIOSH 2010].

outcomes-based, i.e., to compare the LTA prediction with the CDEM prediction and judge both against actual explosion tests conducted with samples using a laboratory test chamber [Cashdollar 1996, Cashdollar and Chatrathi 1993, Cashdollar and Hertzberg 1989, Cashdollar et al. 1987, Cashdollar et al. 1992a, 1992b, and 1992c, Lucci et al. 1995, Weiss et al. 1989]. The following sections detail how this approach was used for this report.

Joint Study between NIOSH and MSHA

A cooperative study between NIOSH and MSHA entailed an MSHA inspector from each of the 10 bituminous coal districts using the CDEM in mines to identify potentially explosible dust mixtures in real time during routine band sample surveys. The study was initiated with one training session held at the National Mine Health and Safety Academy in Beckley, WV (see Appendix B). The mine inspectors were to collect the band sample during their routine survey, measure the explosibility with the calibrated CDEM (per the methods described in Appendix C), and then send the mixed sample to the MSHA National Air and Dust Laboratory for the routine laboratory determination of the percentage of TIC (% TIC). The remainder of the samples would then be sent to NIOSH for parallel testing. The CDEM output obtained by the mine inspectors was shared with NIOSH researchers as well as the corresponding MSHA TIC for each sample.

After the in-mine band samples were collected by the inspectors, the samples were submitted to the MSHA National Air and Dust Laboratory for routine testing to determine % TIC per the MSHA standard sampling protocol and procedure. The MSHA inspectors were directed to record the CDEM readings in the "Location in Mine" column on their Rock Dust Sample Submission Forms.

The inspectors began using the CDEMs in December 2009 (within one month of receiving the training on the proper use of the CDEM). NIOSH requested the inspectors send pure rock dust samples from each mine sampled directly to the NIOSH Office of Mine Safety and Health Research (OMSHR) Pittsburgh location. Once the % TIC was determined, MSHA was asked to send NIOSH the remainder of the dust samples from the surveys in which a CDEM was used by the inspectors, along with the TIC analyses and moisture content data.

A brief look at the dust sample processing time involved in this study reveals a fundamental problem related to the traditional LTA method. Although the average dust sample processing time in this survey was less than 2 weeks, the total elapsed time from the sample collection date until the laboratory informed the MSHA inspector of the results ranged from 1.7 weeks to 5.9 weeks, with an average of 3.6 weeks (Table 1). This period does not include the time elapsed between receipt of the results by the inspector and notification to the mine. During this processing period, the mine entry where the band sample was taken could be deficient in rock dust, thereby representing a potentially unrecognized and unmitigated hazardous condition.

Table 1. Elapsed time between dates that samples were collected and the MSHA inspector was notified of laboratory results

Period	Average	Std. Dev.
Sample collection—results sent to inspector (days)	25.3	8.0
Sample collection—results sent to inspector (weeks)	3.6	1.1
Lab receipt of samples—results sent to inspector (days)	12.9	6.7
Lab receipt of samples—results sent to inspector (weeks)	1.8	1.0

Results and Discussion

The discussion that follows is based on the verifiable data that allowed NIOSH to identify potential issues with the CDEM instrument and procedures that could be improved.

NIOSH received 591 samples from 37 separate surveys performed in 32 different mines in 5 districts. However, NIOSH received pure rock dust samples for CDEM calibration for only 16 of the 32 mines. Therefore, NIOSH researchers were able to verify sample measurements on 297 samples. The set of 297 (~ 50%) of the 591 field study samples which were verifiable are the basis of this study.[14]

The set of 297 field study samples was both random and similar in relation to accuracy and test conditions for the 591 total samples and represented 5 of the 10 bituminous coal districts. Verification of data included the calibration of CDEMs with the local rock dust supply and testing the coal/rock dust sample provided by Mt Hope. Without a local rock dust sample for calibration, no conclusions could be made as to why results differed between the low temperature ash (LTA) analysis and CDEM.

---

[14] In order to ensure that the samples were representative of the dust samples collected from all MSHA coal districts, a one-sample binomial test was conducted. This test allows for the determination of whether a proportion of "successes" on a two-level categorical variable differs from a hypothesized value. The 591 samples and 297 verifiable samples were compared to a database of 50,914 dust samples collected from all MSHA districts in 2010. Prior to the analysis of these data, the dust variable was recoded to either "Green" or "Red" based on an 80% cut point. The null hypothesis that the proportion of "Green" samples in the database of 591 records was equal to the proportion of "Green" samples in the 2010 MSHA dust survey in underground coal mines was not rejected (two-sided p-value = 0.9744). Additionally, the null hypothesis that the proportion of "Green" samples in the database of 297 records was equal to the proportion of "Green" samples in the MSHA database was not rejected (two-sided p-value = 0.1193). Thus the 591 and 297 samples could have come from the 2010 MSHA dust survey and therefore could be considered as representative samples from all MSHA coal districts.
The one-sample binomial test was also used to determine whether the 297 subset of CDEM samples were representative of the 591samples. The null hypothesis that the proportion of "Green" samples in the database of 297 records was equal to the proportion of "Green" samples in the database of 591 records was not rejected (two-sided p-value = 0.1138). Based on the results of this statistical test, the subset of 297 dust measurements could have come from the 591 dust measurements.

Of the 591 samples that the MSHA inspectors tested with the CDEM, 416 were measured as GREEN, 57 were YELLOW, and 118 were RED (Table 2). The GREEN measurements made up 70% of the total, with the remaining 30% containing RED and YELLOW measurements. The RED and YELLOW measurements indicated that the measured Φ was less than the propagation/extinction limit, Φ^*, signaling the need for more rock dust to mitigate a potentially explosible concentration of that dust (as determined by the S_c/S_r ratio in Appendix A). For the purposes of instrument performance assessment and explosibility determinations, both RED and YELLOW results are considered to be RED, i.e. explosible. Of the 297 samples that NIOSH received for evaluation, 73% were GREEN and thus did not require more rock dust to inert (Figure 3).

To quantify the degree of association, or reliability, between the 296 CDEM samples tested by MSHA and verified by NIOSH (Table 2, row 1, column 4), a Kappa statistic (K) was calculated. The Kappa statistic is used when the data are categorical. The data in this analysis were dichotomized as "Green" and "Red." The calculated value of the Kappa coefficient was 0.6868 (95% CI: 0.5923, 0.7814). This value denotes good reproducibility between MSHA and NIOSH data [Rosner 1990].

Table 2. Comparison of CDEM results as determined by separate MSHA and NIOSH testing[15]

CDEM Readings	MSHA-Conducted—Total Survey				NIOSH-Conducted—Verifiable Samples		
	TIC = IC + H_2O	Moisture-free IC = TIC - H_2O	Assuming 8% IC in CD, 80% RD = 81.6% IC	Verifiable samples; assuming 8% IC in CD, 80% RD = 81.6% IC	Moisture-free IC = TIC - H_2O	Assuming 8% IC in CD, 80% RD = 81.6% IC; rounded to 82%	NIOSH LTA verification
Total number of samples	591	591	591	296[16]	297	297	297
< 80% and GREEN	15	19	19	11	2	2	2
# of GREEN	416	416	416	215	222	222	222
> 80% and RED/YELLOW	58	50	37	28	17	11	7
# of RED/YELLOW	175	175	175	81	75	75	75
RED percentage of total # of samples	30%	30%	30%	27%	25%	25%	25%
GREEN percentage of total # of samples	70%	70%	70%	73%	75%	75%	75%
Disagreeing RED percentage of total RED	33%	29%	21%	35%	23%	15%	9%
Disagreeing GREEN percentage of total GREEN	4%	5%	5%	5%	1%	1%	1%
Agreeing RED percentage of total RED	67%	71%	79%	65%	77%	85%	91%
Agreeing GREEN percentage of total GREEN	96%	95%	95%	95%	99%	99%	99%

[15] When verifying the samples, NIOSH measured at least three different full cups of each sample to ensure adequate mixing of the sample.
[16] MSHA did not provide NIOSH with a CDEM measurement on 1 sample. Thus, 296 samples were analyzed using the CDEM by both MSHA and NIOSH.

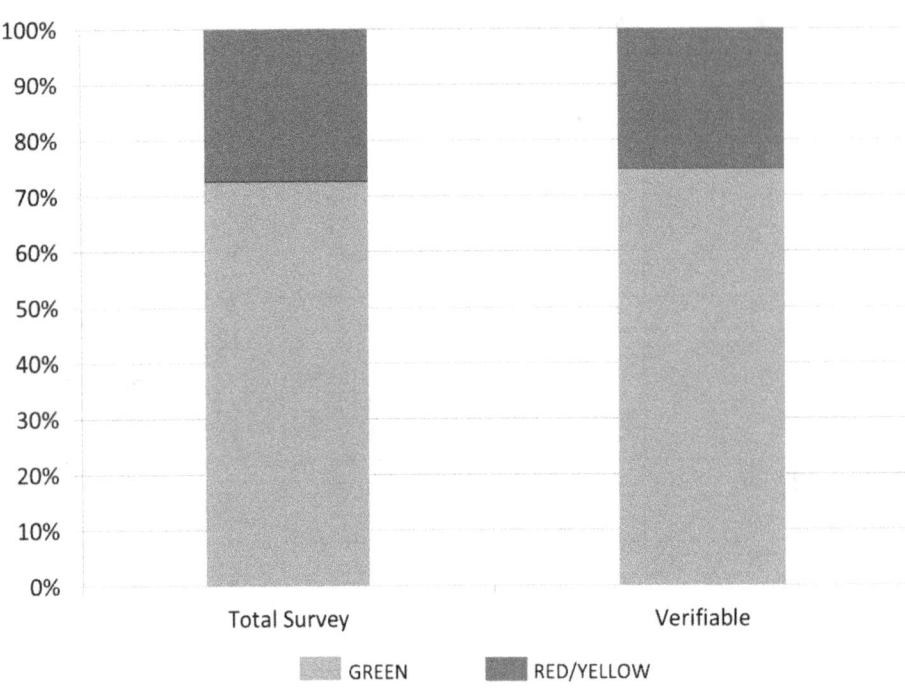

Figure 3. CDEM measurement comparison of NIOSH-received samples.

The samples for measurement with the CDEM needed to be dried prior to determining the reflectance ratio of S_c/S_r. Moisture in a dust sample will darken the sample and thus give an incorrect measurement. MSHA includes the percentage of as-received moisture as part of the TIC. Therefore, in the NIOSH testing, to better compare the sample results from the CDEM to the TIC analysis, the percentage of as-received moisture was subtracted from the % TIC to obtain a moisture-free IC measurement. Subsequently, 19 of the 416 GREEN measurements were < 80% TIC on a moisture-free basis.

All of the GREEN measurements on the CDEM would be > 80% IC if the S_c/S_r were smaller than the S_c/S_r at the extinction limit Φ^*. Twenty-eight of the 57 YELLOW measurements and 22 of the 118 RED measurements were > 80% TIC. All YELLOW and RED measurements on the CDEM should be < 80% TIC if the S_c/S_r is larger than the S_c/S_r at the Φ^* extinction limit.

Because the CDEM and TIC method use different techniques for determining the explosibility of a coal and rock dust mixture, Figure 4 is used to provide a visual comparison of the output results for both methods. Shown in Figure 4 is data collected from one representative dust survey from one mine. The extinction limit, Φ^*, of 0.4 for this mine is indicated on the graph by a horizontal blue line. When the CDEM measured samples with Φs greater than 0.4, the CDEM indicated GREEN (green background). For samples with Φs less than 0.4, RED/YELLOW was indicated (red background). The vertical green line indicates the 80% TIC limit specified for compliance with 30 CFR 75.403.

Based on Figure 4, the following sections interpret the disagreements between the two methods. The terms "disagreement," "agreement," "disagree," and "agree" are used throughout this report to discuss the differences in the CDEM explosibility readings to that of the LTA incombustibility determinations; however, the use of these terms should not be implied as an endorsement of the LTA method as that is not the intent. The focus will be on locations where the points fall in the upper left section of the green background (i.e., the CDEM determined samples to be nonexplosible but the TIC results indicate explosible) and lower right section of the red background (the TIC results indicate nonexplosible but the CDEM indicates explosible).

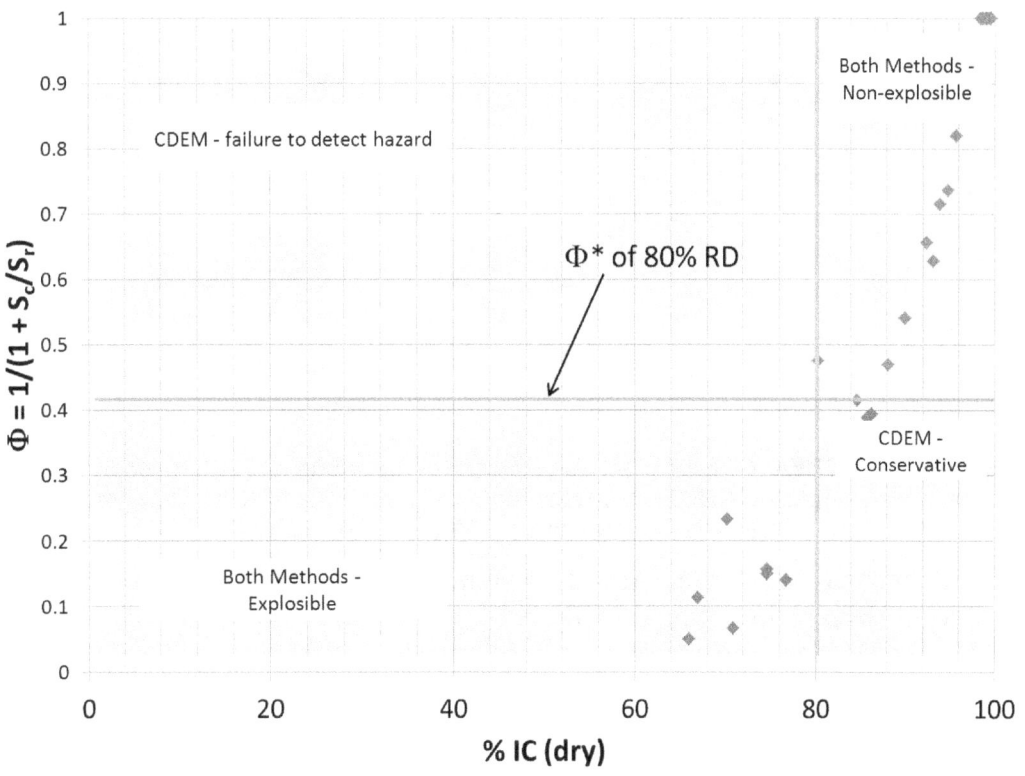

Figure 4. Comparison of CDEM results and lab TIC results for one representative mine.

GREEN Measurements

Of the 416 GREEN measurements documented by MSHA inspections while using the CDEM, 19 (5%) were < 80% IC (< 20 mesh material), which may indicate a failure to detect a hazard (Figure 5). NIOSH was able to identify in 11 of these 19 samples the source of the disagreement. The other 8 samples belonged to surveys and mines for which NIOSH did not have rock dust to calibrate the CDEM.

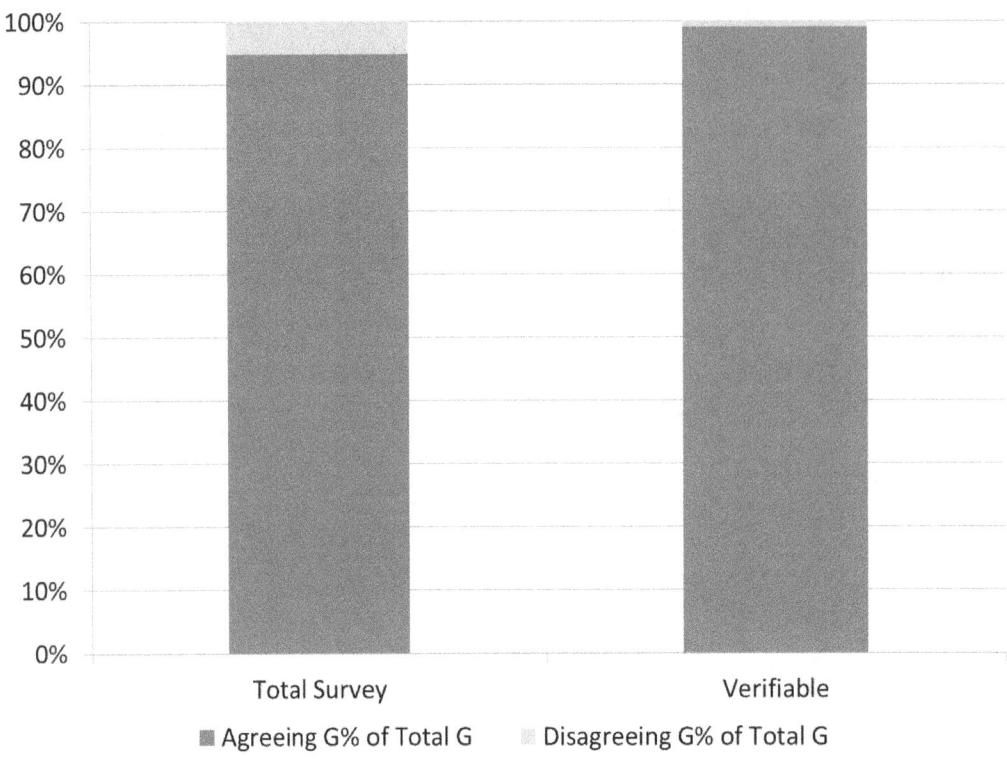

Figure 5. CDEM GREEN (G) measurement comparison of NIOSH-received samples. Agreeing GREEN samples have LTA results > 80% IC. Disagreeing GREEN samples have < 80% IC LTA results.

In 11 verifiable samples (Table 2, row 2, column 4), MSHA determined these to be nonexplosible (GREEN) when using the CDEM but explosible (< 80% IC) with the LTA method. NIOSH CDEM measurements indicated that only 2 (Table 2, row 2, column 5) of these 11 samples were nonexplosible (GREEN). Although the source of the error cannot be confirmed for the other 9 samples, NIOSH researchers believe that simplification of the calibration procedure and more robust training would likely eliminate the variability in results. Upon further examination of the 2 samples, one sample was 79.4% IC and the other was 73.9% IC. These two cases represent < 1% verifiable disagreeing GREEN measurements (Figure 5). Further analysis of these perceived failures to detect was then undertaken.

The MSHA Laboratory where the samples were tested reports a ±0.5% error in its TIC results. The 79.4% IC sample is within this error margin. The 79.4% IC would be explosible only if the particle size of the coal dust were at the finest range (80% < 200 mesh) found in modern underground coal mines. The GREEN nonexplosible result on the CDEM is reasonable considering the likely particle size of the sample presented and the IC reported.

Visual inspection of the 73.9% IC sample indicated larger coal particles as shown in Figure 6. NIOSH measured the percentage of IC (% IC), which does not include moisture, on the < 20 mesh fraction of this sample to verify that the sample received was labeled correctly and matched the MSHA % IC data minus the moisture (i.e., that no clerical error occurred in sample transfer from MSHA to NIOSH). The sample was then subdivided into three particle size ranges to determine the distribution of the inert component within these size ranges. The NIOSH-determined IC was 72.7% IC for the < 20 mesh sample as compared to MSHA's 73.9% IC, showing reasonable agreement given the coarse nature of the sample.

Table 3 shows data for three separate size fractions of the sample. Larger particles in the 20–60 mesh (250–850 μm) size fraction make up 28.5% of the sample mass and contained 52.4% IC. The finest particles in the < 200 mesh range make up 38.7% of the sample mass and contained 86.1% IC. The range from 60–200 mesh particles make up 32.8% of the sample mass and contained 74.9% IC. Experimental studies have shown that the 20–60 mesh (250–850 μm) size fraction does not contribute significantly to flame propagation or inerting. If the sample were normalized for the < 60 mesh portion, the sample would be 81% IC. Further, a GREEN measurement would be expected in this case from the CDEM, because the optical approach is strongly favored by the finer reactive and inerting components. In contrast, the TIC analysis does not distinguish between the larger, nonreactive coal particles and the finer particles for explosibility assessment. Further discussion of particle size effects can be found in Appendix D.

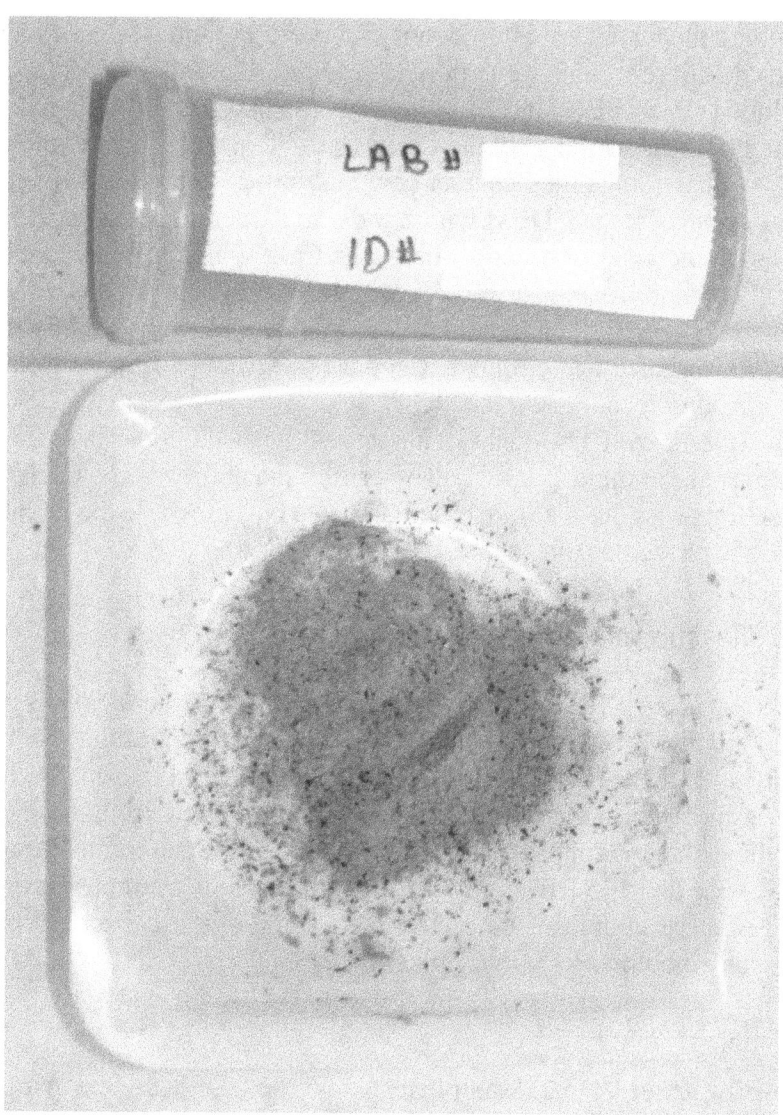

Figure 6. GREEN sample of 73.9% IC.

Table 3. Particle size distribution of 72.7% IC dust sample and LTA results for three different size fractions of the sample

Mesh	Size, μm	Percentage of size fraction	% IC of size fraction
20–60	250–850	28.5	52.4
< 60	< 250	71.5	81.0
60–200	75–250	32.8	74.9
< 200	< 75	38.7	86.1

The NIOSH sample analyses detailed above demonstrated that the measurement discrepancies (where the CDEM did not identify a hazard but the laboratory TIC analysis did identify a hazard) do not represent a true deficit in safety or an instrument inaccuracy. The MSHA data included 19 GREEN CDEM measurements that the TIC data indicated were < 80% IC (Table 2). Potential reasons for why the CDEM provided a different explosibility assessment may be due to the sample containing coarser coal dust, finer inert particles, incomplete mixing of the sample before testing on the CDEM within the mine, operator error, and/or incorrect CDEM calibration. If the sample was not homogeneous, the CDEM field of view could include an area of lighter color where more rock dust was deposited. Researchers are unable at this time to confirm, for these samples, what caused the difference in explosibility assessment in the field study. However, the variance was eliminated in the 11 verifiable samples through proper calibration and operating techniques.

When correctly used, the CDEM shows agreement with TIC in 99% of cases in identifying fully protected areas of the mine. NIOSH also verified the ability of the CDEM to properly assess the explosibility of a dust sample based on particle size where the laboratory result indicated a deficiency based on TIC alone.

RED/YELLOW Measurements

The CDEM RED and YELLOW measurements required further analysis, with a few of the measurements containing greater amounts of IC based on ashing of the < 20 mesh material. Preliminary analysis of these samples identified several potential reasons for the differences in explosibility assessment using the CDEM and the ashing method. Initially, the YELLOW measurement indicated a marginally explosive dust mixture, signaling that more rock dust should be added. MSHA recorded 175 RED and YELLOW CDEM measurements, and TIC analysis indicated that 58 of these samples were > 80% TIC. Some of these discrepancies can be easily explained, as discussed below.

Insufficient Mixing of the Band Sample

It is not known whether the band sample tested on the CDEM by the inspector was the same as that sent to the MSHA laboratory. If the band sample was not sufficiently mixed by the inspector, the portion tested by the CDEM could be different from the portion retained and analyzed by MSHA. It is not uncommon that there are differences in the % IC measured by MSHA and those measured by NIOSH following the same laboratory LTA method. Previous studies by Harris et al. [2008] have shown that there can be ± 7% IC variation in the LTAs if the sample was not adequately mixed after being collected and before being subdivided and tested (Figure 7). In comparison, the reproducibility of the data acquired with the LTA method on well-mixed samples resulted in a standard deviation of 1.7% for moisture-free LTA results from a random set of 29 samples analyzed by MSHA, NIOSH, and an independent certified laboratory.[17]

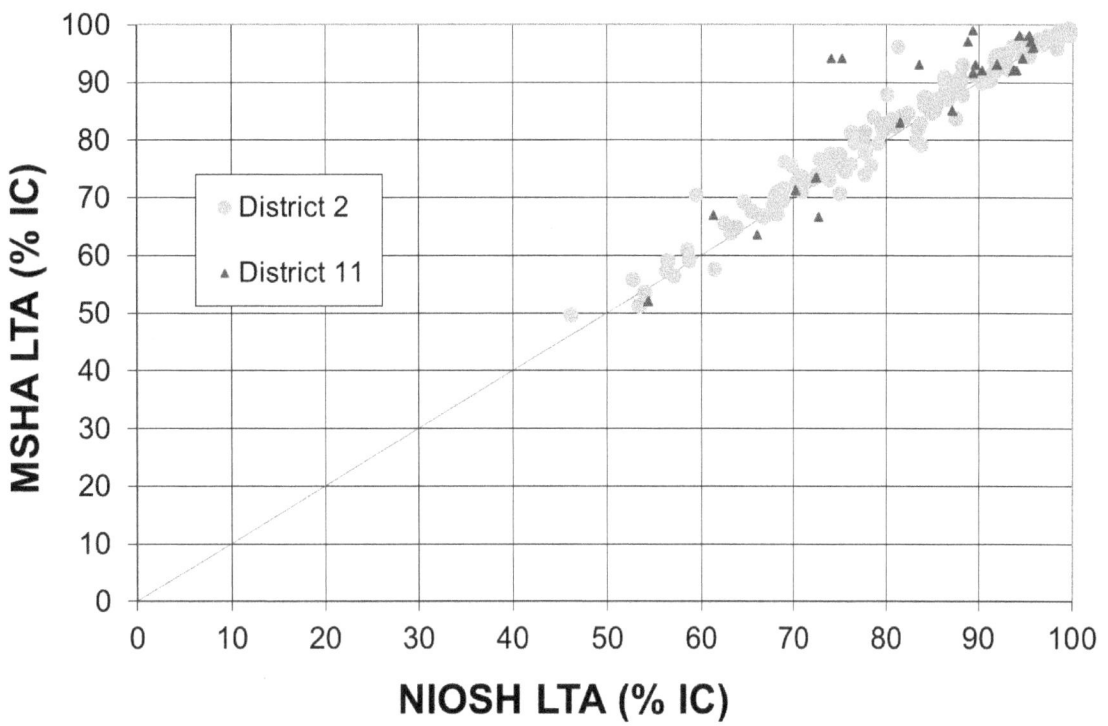

Figure 7. Comparison of MSHA and NIOSH incombustible measurements [Harris et al. 2008].

Insufficient Drying of the Band Sample

The CDEM sample is dried prior to testing. Moisture in the sample will cause the dust mixture to reflect less light and appear to contain less rock dust than is actually contained in the sample. Previous CDEM reflectance studies have shown that for each 1% added moisture, the reflectance equivalent produced is of 7% less rock dust. For example, when 1% water is added to a 65% rock dust/coal dust mixture, the reflectance is equivalent to a mixture containing 58% rock dust [Sapko and Verakis 2006]. Hence, a "false" YELLOW or RED measurement would be

---

[17] To determine the reproducibility of LTA measurements among three independent laboratories (MSHA, NIOSH, and an independent laboratory), 29 well-mixed dust samples were analyzed and an intraclass correlation coefficient (ICC) was computed. The value of the ICC was equal to .982 (95% CI: .960, .992) with an F-test that the true value of the ICC was 0 being rejected at p = .000. The magnitude of the computed ICC shows excellent agreement among the three laboratories [Shrout and Fleiss, 1979].

obtained. Based on the results of the published study, after the moisture analysis was subtracted from the MSHA TIC, there were 50 RED/YELLOW samples (reduced from 58) > 80% IC. This sample number was then further reduced, as described in the next section.

On a related note, insufficient drying of the sample before CDEM measurement in the field would cause the sample to appear artificially dark, and the CDEM would measure the reflected light accordingly. By the time NIOSH received the samples from the MSHA National Air and Dust Laboratory, the samples may have been subjected to additional drying during the shipping process. NIOSH also dried the samples with molecular sieves as part of the study process.

Higher % IC of the Coal Mined

The CDEMs were calibrated with an 80% rock dust mixture for the GREEN/YELLOW boundary. However, coal has a certain percentage of IC material also. If there is an assumed 8% IC in the coal, the 80% rock dust actually contains 81.6% IC. There were 50 MSHA CDEM RED/YELLOW measurements that were above 80% IC (without factoring in moisture). When comparing the dry MSHA IC analysis of 81.6% or higher (assuming 8% IC inherent in the coal) with the CDEM measurements, the number of RED/YELLOW measurements that are > 81.6% IC is 37 individual band samples. By calibrating with a percentage of rock dust (% RD) rather than % IC, a margin of safety is included in the explosibility assessment.

NIOSH had received pure rock dust for calibration of the CDEMs from MSHA inspectors in various districts, but not for all of the mines where MSHA used the CDEM. This limited the number of surveys for which NIOSH could confirm the MSHA inspector's measurements. Out of the 37 surveys, NIOSH could only calibrate the CDEM and verify the results of 16 surveys. Therefore, the individual band samples with RED/YELLOW measurements that were > 81.6% was reduced to 28 samples which NIOSH researchers could confirm (Figure 8). Of these 28 samples, NIOSH was able to replicate RED/YELLOW measurements for 11 of them. TIC results conducted at NIOSH identified 4 of these samples as being < 81.6% TIC, likely indicating sample variance. NIOSH could not replicate RED/YELLOW results for 17 of the 28 samples, indicating problems with calibration or operation of the CDEM. The 11 of the 28 samples that were verified as RED/YELLOW but > 81.6% TIC represent cases where the CDEM errs on the side of safety.

Finally, if the operator is mining material that is dark (such as darker shale), the dust sample may appear darker, yet still contain a higher IC than what the CDEM detects and produce a conservative result.

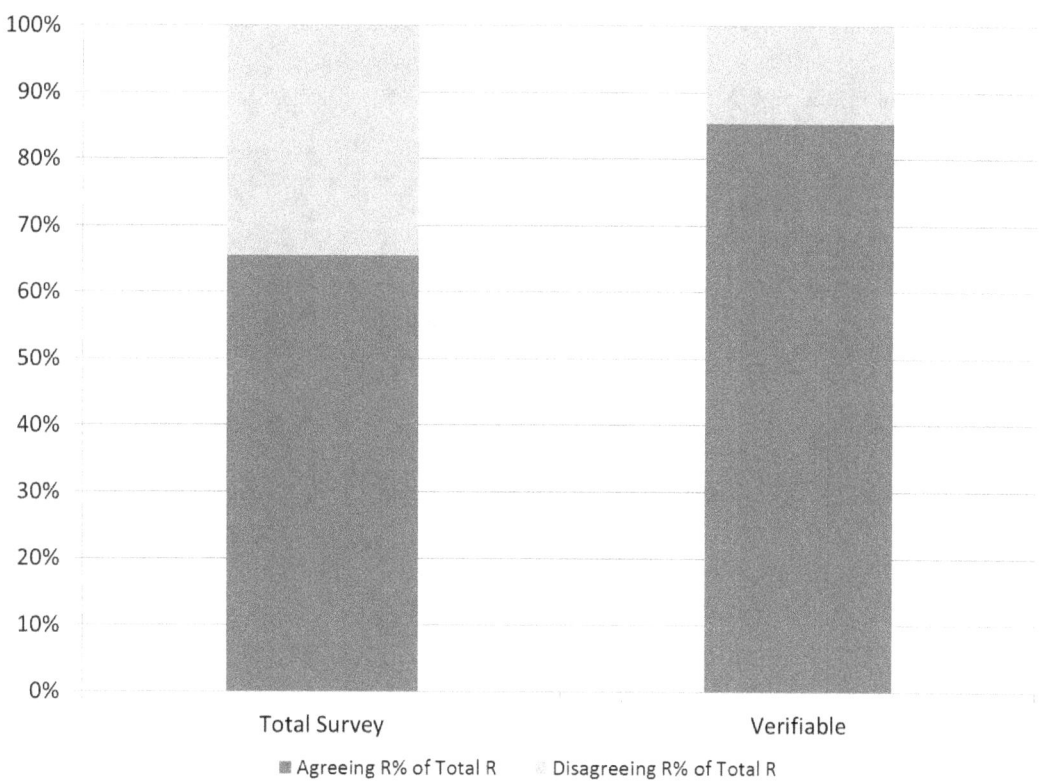

Figure 8. CDEM RED (R) measurement comparison of NIOSH-received samples. Agreeing RED samples have LTA results < 81.5% IC. Disagreeing RED have more than 81.5% IC LTA results.

CDEM Calibration Error

Other sources of error related to the RED/YELLOW measurements could be due to calibration of the CDEM by the user who prepared the calibration samples. If the proportions of rock dust and PPC were not accurately measured out for the 75% and 80% rock dust calibration samples, the set points for the results would not be correct. If the proportions of rock dust and PPC were correct but the calibration samples were not sufficiently mixed, the calibration for color would be affected. In addition, the CDEM was to be calibrated for each mine with the rock dust from that mine before CDEM measurements of the band samples were taken from that mine. If the CDEM had been previously calibrated for use in a different mine, differences would occur if the rock dusts were different.

Of the 28 RED/YELLOW samples that were > 81.6% IC, only 7 were found in the lab to be in disagreement with the TIC results. Of the RED/YELLOW samples measured by NIOSH, 90% of the measurements identified areas that required more rock dust. NIOSH was unable to find an explanation as to why the 7 measurements were at odds with the ashing approach. As with the GREEN cases, there will be some disagreement between the methods because of their fundamental differences. This disagreement represents 7 verifiable cases where the CDEM was more conservative than the laboratory technique. As seen in Figure 9, after accounting for the differences between the two methods, there is 97% agreement between the CDEM and the laboratory analysis.

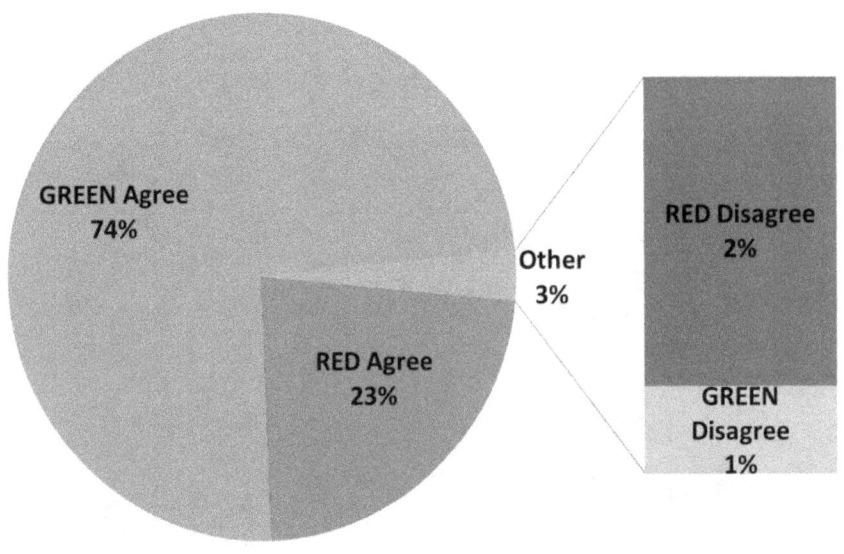

Figure 9. Agreement between CDEM and laboratory analysis.

Conclusions from the NIOSH Study

The CDEM was able to identify (*in situ*) coal and rock dust mixtures that were not capable of supporting flame propagation for 99% of the samples tested, subsequently confirmed by laboratory analysis as containing > 80% IC. The one sample that was determined to be < 80% IC by laboratory analysis was investigated in more detail. After further analysis of this questionable sample, it was found that it contained a large fraction of nonreactive coal and rock dust particles in the range of 20 to 60 mesh (250 to 840 μm). Rock dust particles in this range have been shown to be ineffectual for inerting flame propagation [USBM 1933]. The sample was passed through 60 mesh screen and the < 60 mesh material was ashed to determine the % IC content of the reactive particles. The resulting 81% IC supported the CDEM GREEN reading of a nonexplosible mixture. Therefore, in no case did the CDEM fail to detect an explosion hazard in this study.

Assuming that the CDEM is properly calibrated and the test sample dried, a sample can be considered to be nonexplosible if the CDEM indicates a GREEN measurement; i.e., no additional rock dust is needed at that time. However, it is important to continue to sample that area from time to time because of changing conditions caused by the dynamic nature of the mining environment.

The CDEM and TIC method use two different surrogates for estimating explosibility of coal and rock dust mixtures. Neither measures explosibility directly and therefore neither is indisputable. The only direct measure of explosibility is by use of a 20-L or larger explosion apparatus or through full-scale experiments [Cashdollar 1996, Cashdollar and Chatrathi 1993, Cashdollar and Hertzberg 1989, Cashdollar et al. 1987, Cashdollar et al. 1992a, 1992b, and 1992c]. In the 1950s, the TIC measurement replaced the volumetric method as a means of measuring explosibility. The CDEM reading is weighted by relative coal and rock dust particle sizes in assessing sample explosibility, whereas the LTA method gives equal weight to all dust particles < 20 mesh (840 μm).

Even though the CDEM and the TIC methods for determining the explosibility of coal and rock dust mixtures use different techniques, the results of this field study show excellent agreement (97%) between the two methods. It is unrealistic to expect the CDEM to replicate all laboratory TIC determinations. Rather, the goal is for the CDEM to accurately identify the potential *in situ* explosibility of a dust mixture at the time of sample collection. In this respect, the CDEM can be used to both identify potential hazards in real time and for use as a followup tool to ensure that adequate rock dust was added to abate the potential explosion hazard.

In this study, the CDEM indicated that 25% of the samples collected were potentially explosible. Since each sample in this study represents 500 linear ft of mine development entry, the 175 RED samples represent about 17 miles of mine entries deficient in rock dust that would have been identified as a hazard at the time of the sample collection. The GREEN results, indicating that at the time of testing no additional rock dust is needed to make the sample nonexplosive, were accurate. The RED and YELLOW results, indicating that the area is deficient in rock dust, err on the side of safety in that these results require the addition of more rock dust even if subsequent detailed laboratory analysis indicates otherwise. With the immediate results, the mine can remedy the situation directly rather than wait days or weeks for laboratory results in order to effect change and protect miners.

The CDEM offers an opportunity to immediately identify and correct explosible accumulation of coal and rock dust mixtures. This explosibility assessment is based on the analysis of a representative coal and rock dust mixture collected along the entries. In contrast to the current laboratory analysis method that takes days or weeks to obtain results, the CDEM can immediately identify deficient areas that can support flame propagation and trigger corrective measures to abate the hazard.

Finally, based on the field study results, a need for enhanced training in the use of the CDEM and a simpler method to calibrate the CDEM to minimize errors was identified. This input was used to develop a simpler calibration procedure and improve on-screen instructions to the user in executing calibration and testing procedures.

Commercial CDEM Development

The term CDEM originated with the U.S. Bureau of Mines (USBM) and NIOSH development of the tool. The commercial CDEM is known as the CDEM–1000 as named by the manufacturer. The commercial CDEM developed in 2010–11 is similar to the CDEMs used in the study. However, some changes to the CDEM have been made based on analysis of the study results, input from the inspectors, and input by the instrument manufacturer.

In response to coal mine inspector comments (Appendix E), updates to the CDEM include changes in screen prompts to plain language and more explicit instructions, a simplified calibration process (using three instead of four samples), and consideration of atmospheric methane in results displayed.

The changes in the CDEM screen prompts now lead the user through the use and calibration of the device so that all terms are common and well understood by the average user. This improvement is accompanied by a user manual with thorough descriptions and explanations of the calibration and use process.

Calibration and Programming of the Commercial CDEM

The commercial CDEM–1000 is calibrated using the same standards as those used in this study, except only three samples are required: 100% PPC, 100% RD, and 80% RD. All former YELLOW measurements are now considered to be RED. In the analysis of the study, all YELLOW measurements were considered to be RED readings and thus deemed to require more rock dust. The 80% rock dust sample was chosen as the RED/GREEN boundary based on full-scale experimental results conducted at the NIOSH LLEM [NIOSH 2010]. This value is also reflected in the recent MSHA final rule requiring 80% IC in all entries of an underground coal mine as recommended by NIOSH Report of Investigations (RI) 9679. The changes in the calibration procedure are software changes and not changes to the components or operation principles of the CDEM.

The CDEM programming was also changed to accommodate the presence of methane. In accordance with 30 CFR 75.403, "where methane is present in any ventilating current, the percent of incombustible content of such combined dust shall be increased 0.4 percent for each 0.1 percent of methane." To allow the CDEM to determine the amount of rock dust required in a mine entry, the user is prompted to enter ambient methane levels (as % CH_4) measured with an MSHA-approved methane meter. The CDEM will adjust the RED/GREEN boundary based upon the % CH_4 input, and the requirements of 75.403.

Another change in the CDEM output is to indicate the level of noncompliance to aid mine operators in defining the level of action required to abate a dust explosion hazard. If a sample is indicated as RED, i.e., explosible, the display will indicate a range of % rock dust measured which is based on the measured value rounded up or down to the nearest integer percentage and the rounded value minus 3%. This range does not include the IC inherent in the coal nor the variable as-received moisture of the sample.

Commercial CDEM Changes based on Potential Customer Concerns

Based on the outcomes of this study, an anticipated potential customer concern may be the availability of PPC for calibrating the CDEM. The manufacturer and NIOSH have a PPC supply available to support the initial build of 3,000 instruments. PPC will be provided with each unit for calibration. An evaluation of a manufactured PPC product is currently being pursued to ensure that additional PPC can be produced and supplied to customers to sustain CDEM support indefinitely.

Another potential customer question involves the frequency of calibration. The manufacturer is requiring that the CDEM be recalibrated after 200 measurements. However, if a mine changes its rock dust supplier, the CDEM would need to be recalibrated with new calibrations samples made from the rock dust received from the new supplier. Further, if an operator changes rock dust supplies frequently, the operator will need to recalibrate the CDEM more often. It should be further noted that a highly variable rock dust supply (in color and in particle size) would make visual hazard assessment without a meter extremely difficult and current visual determination of adequate abatement of a violation equally suspect. Only if the current particle size specifications were tightened and if rock dust supplies were certified/controlled would any significant variations from batch to batch and from supplier to supplier be significantly minimized.

The Commercial CDEM as a Verification and Compliance Tool

Discussion of the commercially developed meter would not be complete without identification of potential uses of the instrument in managing dust explosion hazards and compliance with regulations. The CDEM as a compliance assistance tool can be used by mine operators to evaluate accumulations of dust in areas where dust is generated such as returns from producing sections, coal crushing and transfer points, and conveyors or other coal transport systems in entries. The collection and analysis of samples during normal operations will provide an indication of the level of safety provided by a mine's rock dusting program. Areas deficient in rock dust can be identified and improvements to either rock dusting practices or dust control can be enacted. Mine safety inspectors can utilize the CDEM to screen samples for compliance immediately upon collection and take immediate action to have the hazard abated. Samples identified as deficient can then be sent to laboratories for analysis per the appropriate existing practice for determining the potential explosibility of the mixture. Samples identified as adequate may not require laboratory analysis. Finally, should an area be identified as deficient, the CDEM can be used to verify that the hazard was abated in place of the existing qualitative visual assessment.

In addition to these routine applications of the CDEM in traditional dust surveys and spot checks, it can also be used to evaluate compliance with mine ventilation standards. 30 CFR 75.323.d.1.iii requires rock dust to be continuously applied to allow operations up to 1.5% CH_4 in return air splits. The commercial CDEM with the methane input capability can evaluate the adequacy of the effort at rock dusting to ensure that an explosion hazard is not permitted to exist.

NIOSH Recommendations

The CDEM is recommended for use as a compliance tool to allow real-time assessment of coal dust explosion hazards in underground coal mines for the prompt initiation of corrective actions. The following considerations should be used as part of this recommendation:
- If the criteria of 30 CFR 75.403 remains TIC, then the CDEM should be used by the mine operator and the MSHA inspector to determine the onsite need for immediate corrective action. Inspectors would send samples to the laboratory for TIC determination.
- If the use of TIC as the surrogate for explosion hazard enforcement were changed to % RD and if other changes such as inclusion of dust particle size and elimination of the as-received moisture are considered, the CDEM could serve as the regulatory tool to take enforcement actions, and onsite corrective action in the deficient areas could be taken immediately by both the mine operator and the MSHA inspector.

- The sample collected, dried, and tested with the CDEM can be retained for laboratory analysis to independently verify the % IC relevant to current regulatory standards. For any discrepancies between the two methods, explosion chamber tests could be conducted to determine the inherent explosibility of that representative sample.
- Upon receipt of a RED measurement, the area where the sample was collected should be treated with more rock dust in order to inert the explosible conditions present. Rock dust should immediately be applied to the area of entry where the RED sample was collected until a GREEN measurement is obtained.

Acknowledgments

The authors would like to acknowledge H. Erik Sherer, MSHA, Senior Mining Engineer, for coordinating the MSHA study of the CDEM and the invaluable cooperation of the many MSHA field inspectors who calibrated and used the CDEM on collected dust samples. The authors also acknowledge Terry Montgomery, Supervisor Chemist, of the MSHA National Air and Dust Laboratory, Mt. Hope, WV, for providing the dust samples, the incombustible percentage data, and the incombustible analysis procedures. The assistance of Linda McWilliams, Team Leader of the Surveillance Team in the Health Communications, Surveillance and Research Support Branch of OMSHR, was invaluable for providing statistical analysis of the collected data. The authors also recognize the laboratory assistance of all NIOSH OMSHR personnel for their efforts in the CDEM testing of the dust samples and the dust samples particle size testing. The authors would also like to acknowledge the work of Henry Perlee and Ed Hay of H&P Prototype and Harry Verakis, Senior Projects Engineer in MSHA Technical Support, for assisting with the development of the technology that led to the precursor prototype instruments.

References

Cashdollar KL [1996]. Coal dust explosibility. J Loss Prev Process Ind 9(1):65–76.

Cashdollar KL, Chatrathi K [1993]. Minimum explosible dust concentrations measured in 20-L and 1-m^3 chambers. Comb Sci and Tech 87(1–6):157–171.

Cashdollar KL, Hertzberg M [1989]. Laboratory study of rock dust inerting requirements: effects of coal volatility, particle size, and methane addition. In: Proceedings of the 23rd International Conference of Safety in Mines Research Institutes (Washington, DC, September 11–15, 1989). Pittsburgh, PA: U.S. Department of the Interior, Bureau of Mines, pp. 965–977.

Cashdollar KL, Sapko MJ, Weiss ES, Hertzberg M [1987]. Laboratory and mine dust explosion research at the Bureau of Mines. In: Cashdollar KL, Hertzberg M, eds. Industrial dust explosions. West Conshohocken, PA: American Society for Testing and Materials (ASTM), Special Technical Publication (STP) 958, pp. 107–123.

Cashdollar KL, Weiss ES, Greninger NB, Chatrathi K [1992a]. Laboratory and large-scale dust explosion research. Plant/Operations Progress 11(4): 247–255.

Cashdollar KL, Weiss ES, Greninger NB, Chatrathi K [1992b]. Laboratory and large-scale dust explosion research. In: Proceedings of the 26th Annual Loss Prevention Symposium (New Orleans, LA). Paper No. 72C. AIChE Spring National Meeting, 22 pp.

Cashdollar KL, Weiss ES, Greninger NB, Hertzberg M, Sapko MJ [1992c]. Laboratory and large-scale explosion research at the U.S. Bureau of Mines. In: Proceedings of The First World Seminar on the Explosion Phenomenon and on the Application of Explosion Protection Techniques in Practice (Brussels, Belgium). Section 1, 16 pp.

CFR. Code of Federal Regulations. Washington, DC: U.S. Government Printing Office. Office of the Federal Register.

Harris ML, Sapko MJ, Cashdollar KL, Verakis HC [2008]. Field evaluation of the coal dust explosibility meter (CDEM). Min Engr 60(10):50–54.

Litton CD, Chaiken RF [1996]. The role of radiation absorption in defining explosibility of coal/rock dust mixtures. In: Proceedings of the Twenty-Sixth Symposium (International) on Combustion. Pittsburgh, PA, 1996. pp. 1571–1577.

Lucci CE, Cashdollar KL, Sapko MJ [1995]. Coal dust explosibility meter, Proceedings of the 26th International Conference of Safety in Mines Research Institutes, Katowice, Poland, Sept.4–8, 1995.

MSHA [2010]. MSHA to publish emergency temporary standard for rock dust [http://www.msha.gov/MEDIA/PRESS/2010/NR100921.asp.] Date accessed: 12/01/2010.

NIOSH [2006]. Float coal dust explosion hazards. U.S. Department of Health and Human Services, Centers for Disease Control and Prevention, National Institute for Occupational Safety and Health, DHHS (NIOSH) Publication No. 2006–125.

NIOSH [2010]. Recommendations for a new rock dusting standard to prevent coal dust explosions in intake airways. By Cashdollar KL, Sapko MJ, Weiss ES, Harris ML, Man CK, Harteis SP, Green GM: U.S. Department of Health and Human Services, Centers for Disease Control and Prevention, National Institute for Occupational Safety and Health, DHHS (NIOSH) Publication No. 2010–151.

Rosner B [1990]. Fundamentals of biostatistics, third edition. Boston: PWS-Kent, 655 pp.

Sapko MJ, Verakis H [2006]. Technical development of the coal dust explosibility meter. In Proceedings of the SME 2006 Annual Meeting. St. Louis, MO: Preprint 06–044.

Sapko MJ, Weiss ES, Watson RW [1987]. Explosibility of float coal dust distributed over a coal-rock dust substratum. In: Proceedings of the 22nd International Conference of Safety in Mines Research Institutes. Beijing, China: China Coal Industry Publishing House, pp. 459–468.

Shrout PE, Fleiss JL [1979]. Intraclass correlations: uses in assessing rater reliability. Psych Bulletin *86*(2):420–428.

USBM [1933]. Explosion tests of Pittsburgh coal dust in the experimental mine 1925 to 1932, inclusive. By Rice GS, Greenwald HP, Howarth HC. Bureau of Mines Bulletin 369, United States Government Printing Office, Washington, DC, p. 10.

Weiss ES, Greninger NB, Sapko MJ [1989]. Recent results of dust explosion studies at the Lake Lynn Experimental Mine. In: Proceedings of the 23rd International Conference of Safety in Mines Research Institutes (Washington, DC, September 11–15, 1989). Pittsburgh, PA: U.S. Department of the Interior, Bureau of Mines, pp. 843–856.

APPENDIX A: CDEM Design

The CDEM (Figure A1) consists of an optical probe connected to a small electronics box with a digital display. The optical probe consists of a near-infrared source and a silicon photodiode sensor. When the CDEM probe is inserted into the dust mixture, the near-infrared radiation reflects off the surface of the dust and back to the silicon photodiode sensor.

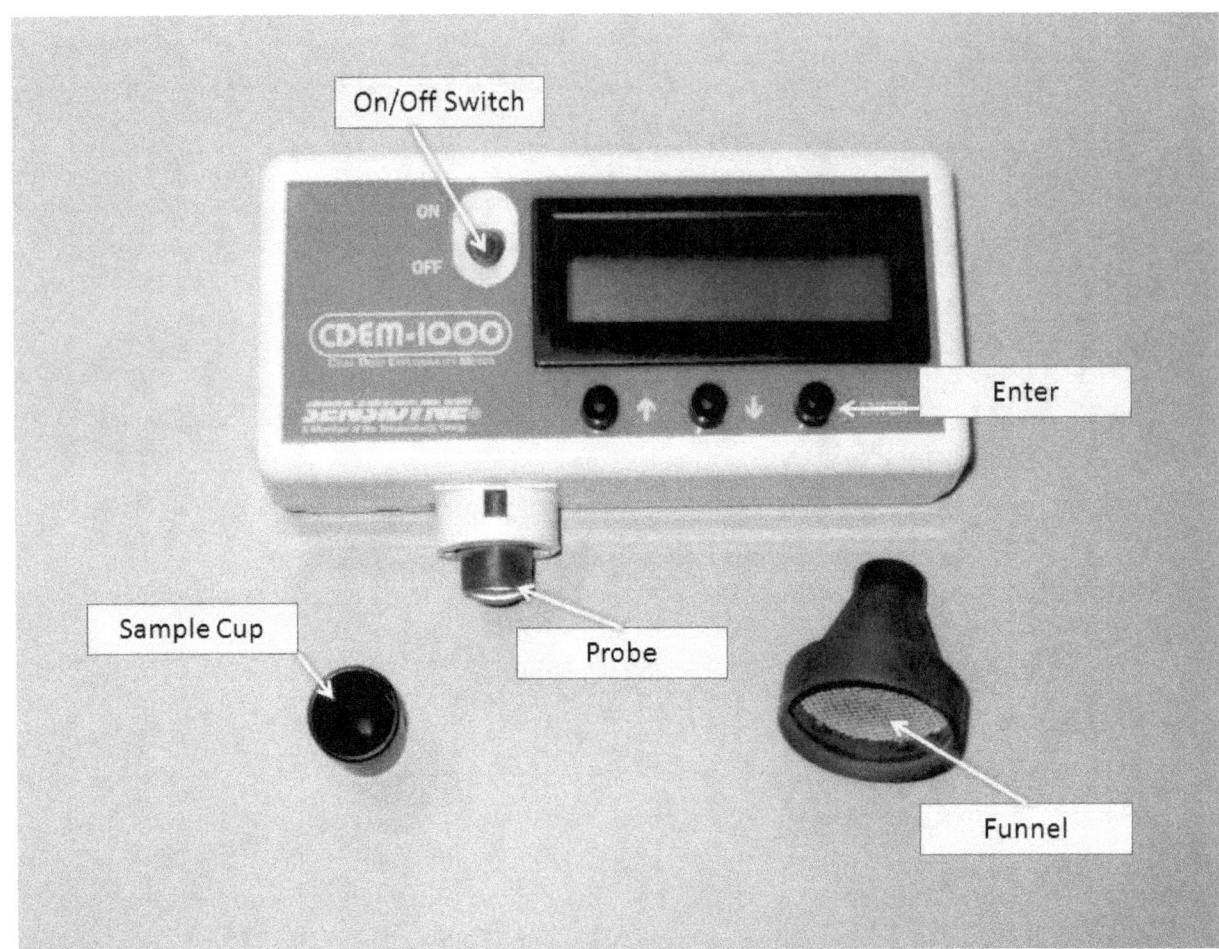

Figure A1. Coal Dust Explosibility Meter (CDEM).

The normalized reflectance, Φ, is related to the mass fraction of rock dust, f_r, in the sample by the following equation [Sapko and Verakis 2006]:

$$\Phi = (Ix - Ic)/(Ir - Ic) = fr/(fr + K(1 - fr)) \tag{1}$$

where I_x = intensity of light reflected from a homogeneous coal and rock dust mixture;
I_c = intensity of light reflected from a pure coal dust surface;
I_r = intensity of light reflected from a pure rock dust surface; and
K = proportional to the product of the rock dust to coal dust particle density ratio and the ratio of the mean particle diameters of coal to rock dust contained in the mixture.

Φ can also be written directly as:

$$\Phi = 1/(1 + S_c/S_r) \quad (2)$$

where S_c = coal dust surface area and
       S_r = rock dust surface area.

For a fixed ratio of S_c to S_r, the quantity Φ is independent of explicit dependence upon K or f_r. This is because the ratio S_c/S_r can be maintained constant while f_r and K can vary.

While examining the data from many coal and rock dust mixtures, the normalized reflectance, Φ, of the dust mixtures near the boundary which separates mixtures that propagate flame and those that do not propagate flame was relatively constant, Φ*, the normalized reflectance at the extinction boundary. To visualize this finding, mixtures were made using various experimental coal sizes that contained 20, 40, and 80% < 200 mesh coal particles used in the full-scale explosion experiments shown in Figure 1. Since the experimental measured rock dust inerting limit has a precision of approximately ± 3 weight (wt) percent, rock dust and coal dust samples were also prepared at ± 3 wt percent around the experimental boundary, and their normalized reflectance was also measured with the CDEM. Figure A2 shows the Φ values measured at the inerting limit for the Pittsburgh seam coal dust sizes shown in Figure 1. The error bars represent the measured Φ of the samples containing ± 3 wt percent rock dust around the boundary between propagation and nonpropagation. As shown, for the Pittsburgh seam coal, Φ at the limit of explosibility is fairly constant over a wide range of volume median coal particle sizes, ranging from 57 μm (80% < 200 mesh) to 270 μm (20% < 200 mesh), while the rock dust decreased from 80 to 68%, respectively [NIOSH 2010]. A larger value of Φ (greater than Φ*) is produced by a smaller ratio S_c/S_r. This would result from an increase in the rock dust surface area and/or a decrease in the coal dust surface area. A smaller value of Φ (less than Φ*) is produced by an increase in the ratio S_c/S_r, and would result in a greater increase in S_c than in S_r. That is, a decrease in S_c/S_r moves the mixture into the nonexplosible range, whereas an increase in S_c/S_r moves the mixture into the explosible range.

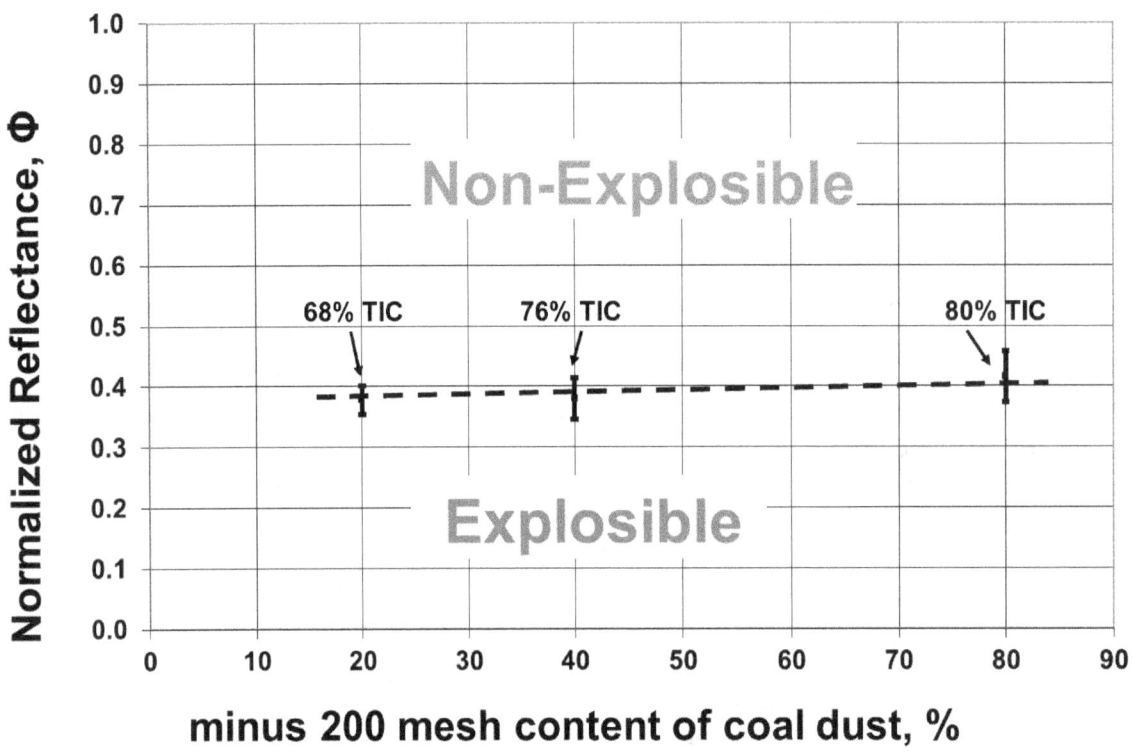

Figure A2. Extinction limit for normalized reflectance [Sapko and Verakis 2006].

The constant, Φ, at the extinction boundary, is also supported by a theory developed by Litton and Chaiken [1996]. This theory observed that, at the lower explosion limit, the product of specific absorption (absorption = 1 - reflection) and mass concentration of coal dust is constant for fixed coal volatility. Several historical coal samples from various coal seams where volatility varied from 15% to 42% indicated that the reflectance of the pure coal was not significantly different from the PPC. Theory and data both support the experimental findings that the CDEM can be used to determine the explosion potential of a particular coal and rock dust mixture.

When calibrating the CDEM, Φ^*, the normalized reflectance at the extinction boundary is set by the mixture made and used as the 80% mine rock dust and 20% PPC.

APPENDIX B: CDEM Training

MSHA inspectors from the 10 bituminous coal districts were trained by NIOSH representatives on the calibration and use of the CDEM at the National Mine Health and Safety Academy in Beckley, WV, on November 5, 2009. Two presentations were given detailing the background and development of the CDEM and the steps to calibrate and use the device. The training presentations were subsequently sent to Erik Sherer, Senior Mining Engineer for MSHA Coal Mine Safety and Health, for later distribution among the inspectors.

Calibration samples were prepared by the inspectors using PPC and the mine's RD in the following percentages by weight: 0% RD (100% PPC), 75% RD, 80% RD, and 100% RD. The 30-g or 50-g sample mixture of 75% and 80% rock dust were to be prepared as listed in Table B1. The 75% RD sample is used to set the boundary between RED and YELLOW. The 80% RD mixture is used to set the GREEN/YELLOW boundary. Calibration samples were required for each mine. If more than one mine used the same RD supplier, the calibration samples may be the same for those mines.

Table B1. Mixtures of RD and PPC for CDEM calibration

Based on 30-g mix				
% RD	Percentage of PPC	RD (g)	PPC (g)	total (g)
80.0	20.0	24.00	6.00	30
75.0	25.0	22.50	7.50	30
Based on 50-g mix				
% RD	Percentage of PPC	RD (g)	PPC (g)	total (g)
80.0	20.0	40.00	10.00	50
75.0	25.0	37.50	12.50	50

MSHA supplied the meters, scales, molecular sieves, and sampling tubes so that the inspectors could prepare the calibration samples for the mine surveyed and collect dust samples to test with the CDEM within the mine. The inspectors were asked to collect a pure rock dust sample from each mine, with some of the sample kept by the inspector in order to mix the calibration samples for each mine and the remainder of the sample sent to NIOSH. NIOSH would then use the pure rock dust samples to make its own calibration samples for subsequent testing and verification of inspector-collected samples after MSHA analysis.

APPENDIX C: Prototype CDEM Calibration and Testing Procedures Used in the Joint Study

Prior to use, the CDEM was calibrated using samples of that mine's pure rock dust, PPC, and two mixtures of known rock dust and PPC content. The following sections describe the instructions provided to MSHA on the proper preparation of the known calibration mixtures, calibration of the CDEM, collection of a representative dust sample, and the use of the CDEM to determine the explosibility of an unknown dust sample.

The following paragraphs describe the calibration and testing directions used by MSHA in this joint study. The procedures for the use of the commercialized CDEM have since changed. The CDEM used for this study was a prototype meter that the manufacturer has since enhanced in the commercialized version. The calibration procedure, meter displays, and operation of the commercialized meter have been simplified for ease of use. Therefore, all of the manufacturer's calibration and use instructions provided with the commercialized CDEM must be followed.

Preparation of Calibration Sample

One sample is used to set the GREEN/YELLOW boundary and a second sample is used to set the YELLOW/RED boundary. The particle size of PPC is characterized in Table C1. For these calibrations, an 80% rock dust mixture was used to set the GREEN/YELLOW boundary and a 75% rock dust mixture was used to set the YELLOW/RED boundary.

Table C1. Pittsburgh pulverized coal size data

Mesh Size	Size Range, μm	Cumulative %
50 x 40	300–425	100
70 x 50	212–300	100
100 x 70	150–212	99
140 x 100	106–150	94
200 x 140	75–106	78
270 x 200	53–75	57
400 x 270	38–53	37
500 x 400	25–38	24
635 x 500	20–25	12

The coal dust (CD) samples used to calibrate the CDEMs are PPC samples supplied by NIOSH. For proper calibration, both the PPC and mine rock dust (RD) should be dried prior to making the boundary calibration samples. The rock dust samples as received for each mine are dried using molecular sieves.

The term "color sample" refers to a dust sample consisting of CD and RD to determine the boundary between RED/YELLOW and YELLOW/GREEN. Color samples are made by mixing RD and CD with the percentage determined by weight—e.g., an 80% RD sample is made by mixing 40 grams (g) of RD with 10 g of CD (Table C2). This mixture does not account for the incombustible content within the coal.

Table C2. Reference mixtures, percentage by weight

Based on 50-gram mix				
Percentage of RD	Percentage of CD	RD (g)	CD (g)	Total (g)
85	15	42.50	7.50	50
82	18	41.00	9.00	50
80	20	40.00	10.00	50
77	23	38.50	11.50	50
75	25	37.50	12.50	50
73	27	36.50	13.50	50

Drying the sample

Each sample must be dried prior to applying it to the CDEM. If the sample is not dried first, the moisture contained in the sample may make the sample appear artificially dark and the CDEM assessment may therefore be incorrect. The following procedure is used to ensure a dry sample:

- Tubes (1½ ounces, oz, or 45 milliliters, mL) half-filled with predried molecular sieves are used to collect the samples (Figure C1).
- The sample tube filled with molecular sieves and test material (either CD, RD, or a color sample) is shaken so that the sample mixes with the sieves and is allowed to dry for at least one minute.

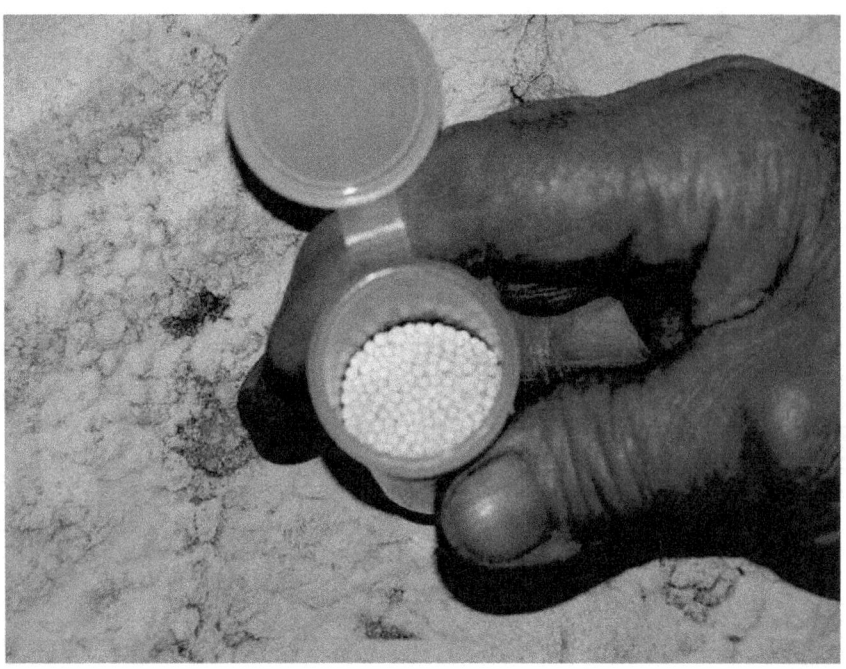

Figure C1. Sample tube containing molecular sieves.

CDEM Calibration

The CDEM must be calibrated for the specific mine before the CDEM is used to determine the IC of the dust samples from that mine. If not, the CDEM will operate based on the previous calibration which may or may not be set correctly for the mine's rock dust used. Therefore, it is advisable to keep a calibration log for each CDEM with the date and rock dust used to calibrate that CDEM for a particular mine. A 50-g mixture is sufficient for calibration.

By preparing 80% and 75% RD mixtures, the chosen color boundaries are GREEN/YELLOW at 80% RD and YELLOW/RED at 75% RD.

The following procedure is used to calibrate the CDEM:
- Turn the CDEM on while pressing the **Enter** button; the meter will then display "Meter Calibration."
- Fill sample cup halfway with dried RD. At RD prompt display, press cup firmly onto probe, and press **Enter** on the CDEM.
- Clean the probe and cup.
- Fill sample cup halfway with dried CD. At CD prompt display, press cup firmly onto probe, and press **Enter**.
- Clean the probe and cup.
- When prompted for GREEN/YELLOW, fill sample cup with the 80% mix, press the cup firmly onto probe, and press **Enter**.
- Clean the probe and cup.
- When prompted for YELLOW/RED, fill sample cup with the 75% mix, press the cup firmly onto probe, and press **Enter**.
- Clean the probe and cup.

- If the calibration was successful, the calibration is saved and the CDEM will display "Calibration complete." At this point, the CDEM can be turned off or left on. If the CDEM is left on, it will proceed to the testing mode.

As indicated by the above procedure, it is important to clean the probe and cup between samples.

In-Mine Testing

The reliability of the CDEM reading is most dependent on the collection of a representative sample using the MSHA band sampling equipment and procedures, as follows:
- Collection of sample (general outline of the dust collection procedures).
    o In the US, routine band surveys are performed quarterly at each mine by MSHA.[18] Samples (up to 1 inch deep) are collected from the floor, ribs, and roof (when possible). Samples are taken every 500 ft in the sections of new development.
    o The MSHA inspector collects the band sample by brushing dust from the ribs, floor, and roof (when possible) into a metal pan (Figure C2).

Figure C2. Collecting a band sample.

---

[18] See pp. 60–66 of the *General Coal Mine Inspection Procedures and Inspection Tracking System* for detailed MSHA procedures. This publication is available at
http://www.msha.gov/readroom/handbook/PH08-V-1GeneralCoalInspectionHandbook.pdf

- The dust is then sifted through a 10 mesh (1.7 millimeter, mm) screen to remove particles larger than 2 mm (Figure C3).

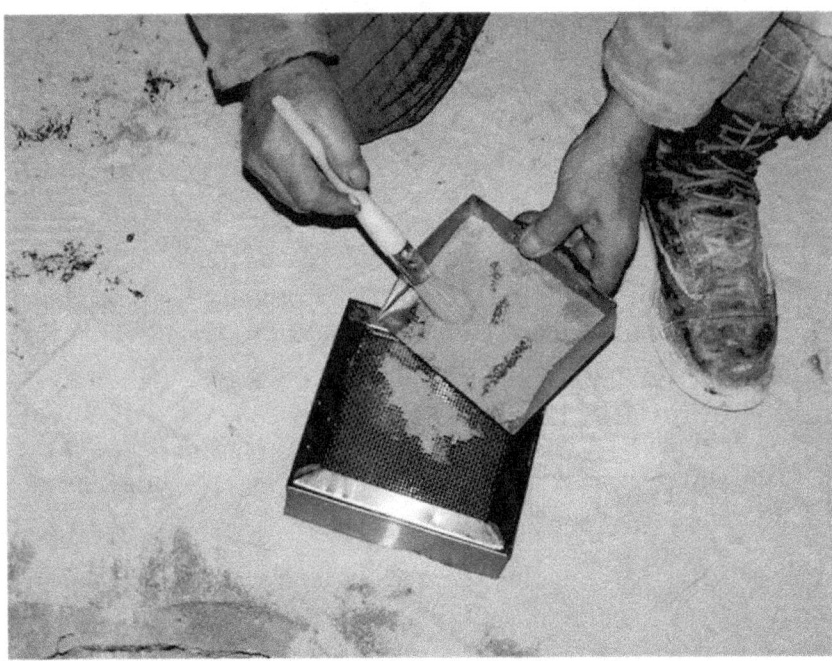

Figure C3. Sifting a sample.

- Once sifted, if there is sufficient dust for a complete sample, the dust is mixed. If not, additional dust should be collected until a sufficient sample is obtained.
- To mix the sample dust, the inspector is to cut, cone, and quarter the mixture to obtain a homogeneous mixture.
- Collecting a sample for the CDEM.
  - Tubes (1½-oz, or 45-mL) half-filled with molecular sieves are used to collect the sample from the tray and to dry the dust (Figure C1 and Figure C4). It is advisable to prepare the tubes by filling them with molecular sieves before entering the mine.
  - The sample tube filled with molecular sieves and dust is shaken so that the dust sample mixes with the sieves and is allowed to dry for at least one minute.

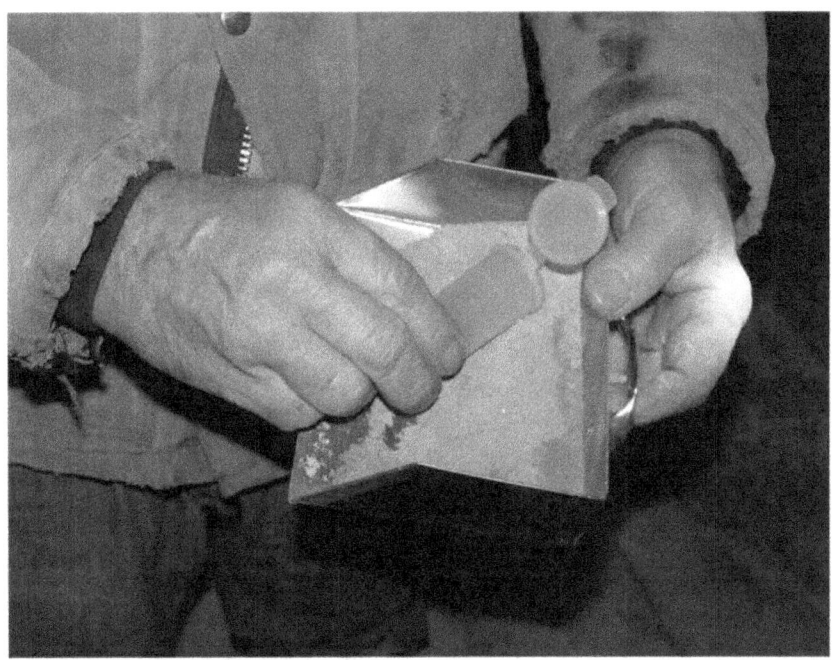

Figure C4. Collecting a sample into a sample tube.

- With the moisture now removed from the dust sample, a funnel with a 20 mesh screen (850 µm) is attached to the end of the tube.
- The CDEM sample cup is fitted onto the end of the funnel and the dust sample is shaken from the sample tube into the sample cup (Figure C5).
- The 20 mesh screen inside the funnel prevents the molecular sieves from entering the sample cup.

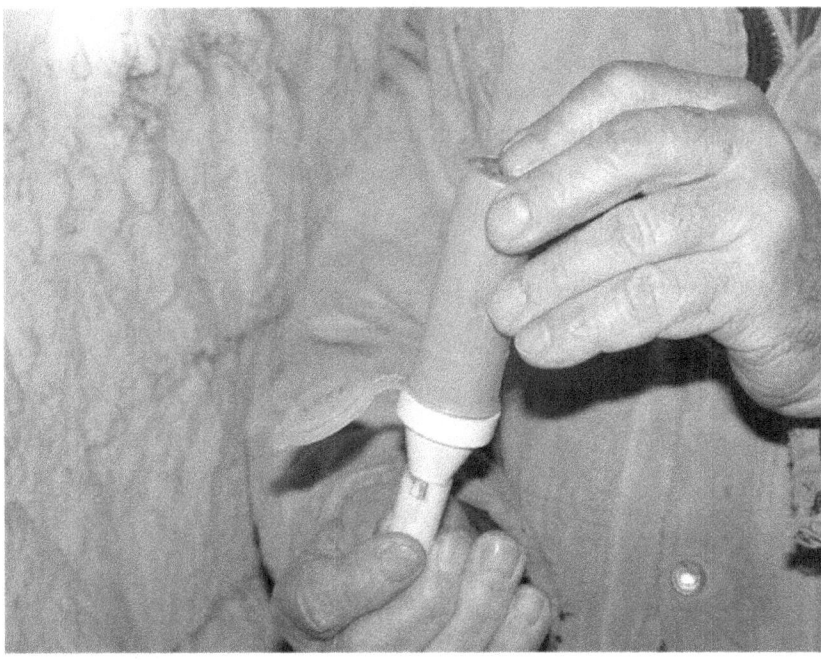

Figure C5. Transferring dust to a sample cup.

- Testing Mode: Testing a mine dust sample with the CDEM after calibration.
  - Turn on the CDEM.
  - When the CDEM is ready, "PUT PROBE IN SUT THEN PRESS ENTER" will be displayed. SUT stands for "sample under test."
  - Once the dry dust is in the sample cup, insert the CDEM probe into the sample cup and hold firmly to the probe lens. While holding sample against the probe lens, push the **Enter** button on the CDEM (Figure C6).

Figure C6. Testing a dust sample with the CDEM.

  - The CDEM returns a reading of explosibility. GREEN indicates there is sufficient RD in the sample presented. YELLOW indicates the sample is marginally explosible and more RD should be added to that area of the mine. RED indicates deficient RD in the sample, requiring more RD be applied to the area of the mine.
- Between-sample cleaning of the CDEM.
  - Gently tap the side of the CDEM against your palm to remove dust from around the probe lens.
  - Using the bare palm of your clean hand, wipe off any dust on the probe lens. Do not use paper or other materials as it could scratch the lens or statically charge the lens.

APPENDIX D: Particle Size Effect

Finer coal particles are more explosible than coarse particles, and finer rock dust particles are better at suppressing an explosion than coarse particles. The reflectance of the coal mine dust responds in a similar manner. The CDEM measures this reflectance to determine explosibility of a dust sample. The CDEM is also sensitive to particle size. The ability of the CDEM to detect variations in particle size in relation to explosibility (see Equations 1 and 2 in Appendix A) is one reason that CDEM results may vary from laboratory IC measurements.

Figure D1 shows the change in the amount of rock dust required to inert three different particle sizes of coal dust. The horizontal line expressing the explosibility threshold was determined by NIOSH testing. Coal dust with a 57-μm mean diameter, the size of the PPC used in NIOSH testing and the size provided as a calibration standard with the CDEM, requires 80% rock dust. Coal dust 10% finer requires about 82% rock dust, while coal dust 10% coarser requires only about 78% rock dust.

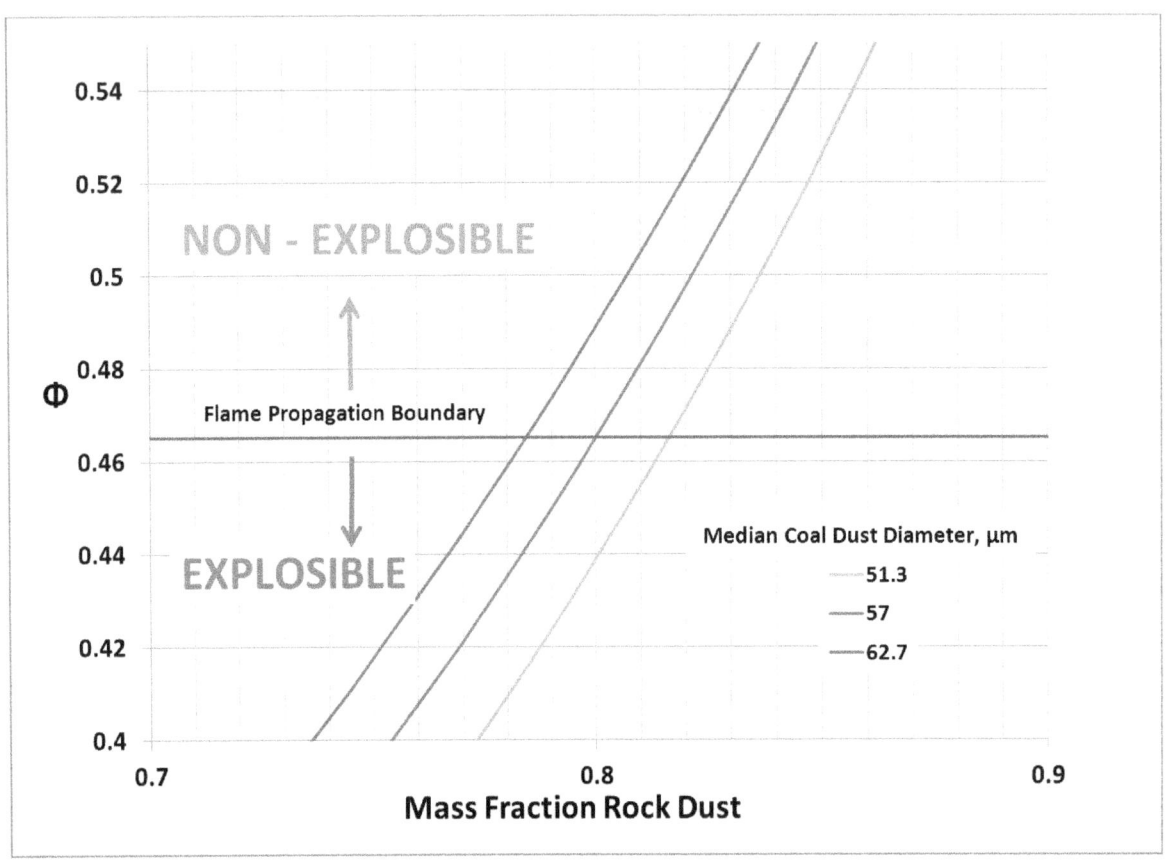

Figure D1. A comparison of coal dust particle size distributions on ϕ values. Rock dust particle size distributions are constant.

Figure D2 shows the same relationship as Figure D1 for variations in rock dust particle size. The nominal rock dust particle size used in full-scale explosion experiments by NIOSH [NIOSH 2010] was 26 µm and should make up 80% of the mine dust for that dust to be considered inert. A 10% increase in rock dust particle size would require almost 82% rock dust to be inert, while a 10% decrease in rock dust particle size would require 78% rock dust to be inert.

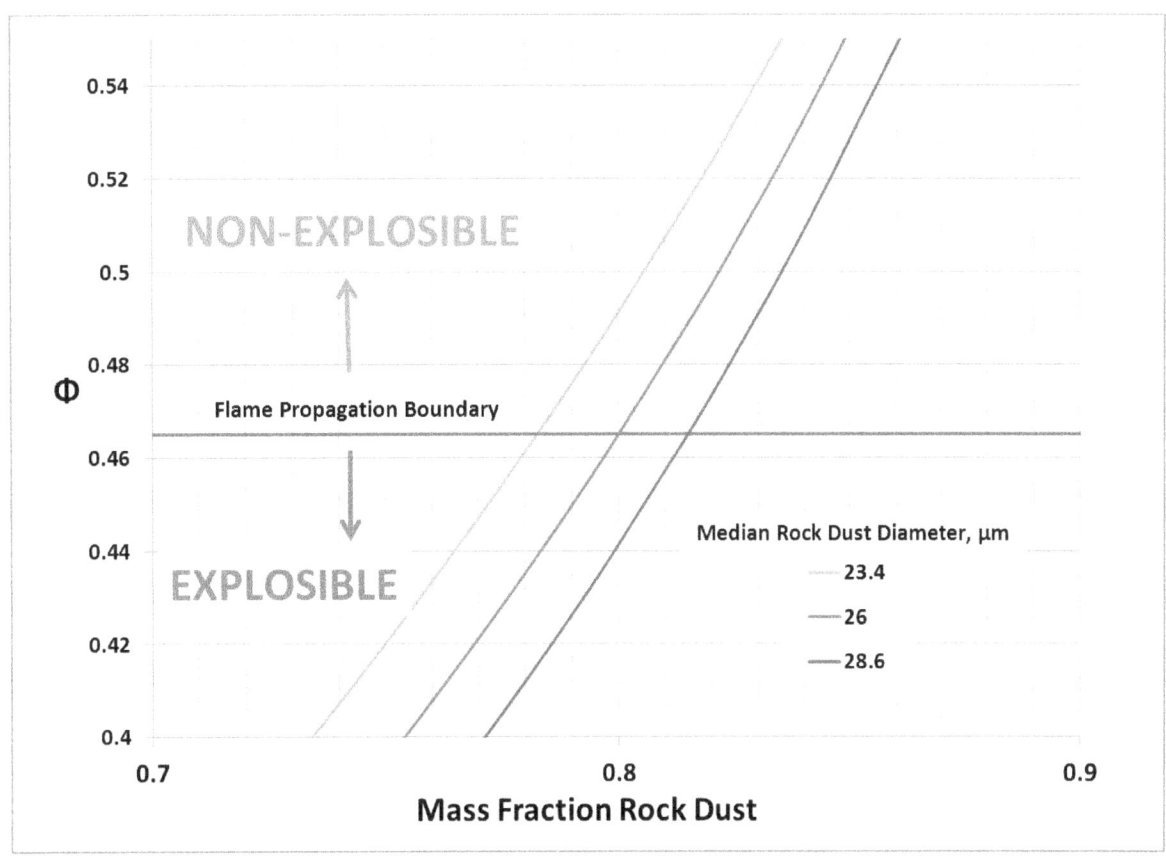

Figure D2. A comparison of rock dust particle size distributions on ϕ values. Coal dust particle size distributions are constant.

If a mine dust contains coal dust finer than the PPC and/or rock dust coarser than a median 26-µm particle size, the CDEM may indicate explosibility (RED) even though the dust is > 80% IC. This result is not an error but an indication that the dust is more explosible than measuring IC alone can determine.

If a mine dust contains coal dust coarser than the PPC and/or rock dust finer than a median 26-µm particle size; the CDEM may indicate nonexplosibility (GREEN) even though the dust is < 80% incombustible content. This result is not an error but an indication that the dust is less explosible than measuring IC alone can determine.

APPENDIX E: MSHA Inspector Questions and Comments

As a followup to this study and to evaluate the use of the CDEM, MSHA asked its coal mine inspectors who used the CDEM in the field four questions regarding the use of the meter:

1. Did you have any problems using the meter?
2. What would you like to change or improve on the meter?
3. Do you think that this meter would be useful for rock dust sampling and enforcement (considering that we will have a regulation change to support the RED and GREEN output)? and
4. Any other comments or suggestions for this meter or on our rock dust program in general?

The boldfaced questions are listed below, with responses in plain text, followed by NIOSH commentary in italicized text. Note that these were the original NIOSH responses to MSHA from each of the questions asked by the inspectors. Based on this study, subsequent enhancements to the prototype meter were since incorporated into the commercialized version of the CDEM.

1. Did you have any problems using the meter?

a. One day the meter would turn off as soon as it was turned on.

Response: NIOSH is unable to determine a cause for this intermittent problem.

b. Hard to use in low coal.

Response: We understand everything done in low coal is more difficult.

c. The drying process requires you to prep a canister for each survey point. This makes larger surveys very burdensome, especially in lower coal seams and where rides are not available.

Response: We understand that all activities conducted in low coal are more difficult to perform and adding this additional assignment of conducting a parallel evaluation using the CDEM at the same time following the traditional sample handling added to the burden. However, it needed to be done as part of this field study in order for MSHA to make additional comparisons between the two methods for assessing explosibility. In the development, implementation, and use of the CDEM, NIOSH researchers envisioned, for purposes of compliance, only sending back the dust sample (small tube with molecular sieves) that were found to be deficient (RED) at the mine to Mt. Hope for followup on regulatory action, if needed. One goal of NIOSH is to quickly identify and correct potential deficiencies in rock dust preferably through the CDEM use by the mine operator.

d. Difficult time reading the information. (This was the most common complaint and made by every inspector.)

Response: The output is either RED or GREEN on the display of the CDEM commercial version. The LCD readability has been improved.

    e. Carrying the tubes for drying the dust is extra weight and time consuming to use.

Response: As highlighted in the response to 1c above, in order to make comparisons, traditional sample collection and handling was required and added more weight to carry in and out of the mine. NIOSH researchers envision through the implementation and use of the CDEM, the total weight of samples carried out of the mine for subsequent laboratory processing will be much less than what is now done as part of the current sample processing procedures.

2. **What would you like to change or improve on the meter?**

    a. The meter needs a back light so it can be easier seen in the dark.

Response: The meter could use this as a feature in a future model.

    b. Would like to see a percentage read out as well.

Response: We do not support this. The percentage readout would only be a prediction and may divert the focus of safety. We recommend the CDEM's use as a compliance check.

The output reading would only be accurate if the median particle size of the rock dust and of the coal dust in the sample were identical to those used in the 80% calibration sample. Since the particle size in the actual samples can vary and the < 20 mesh (841μm) dust fraction may contain very large ineffective inert particles (> 250 μm and < 841 μm), outputting the % IC is of little value and may not be subsequently confirmed by conventional LTA analysis and thus generate more confusion.[19]

    c. Develop a way to roll the meter along the rib, ground, and roof instead of sweeping a sample.

Response: Interesting idea and conceptually has merit. However, at this time, that is not possible with this technology. Such an approach would only be feasible if one could assure that the scanned surfaces were dry and the dispersible surface coal and rock dust is not stratified but well mixed.

---

[19] Due to several requests such as this, the commercial version has the option to display a range of predicted percentages of rock dust if the measurement is RED.

d. Line or lip for testing cup.

Response: Good suggestion, the manufacturer has included a "fill line" in the newly designed sample cup.

e. The calibration process is time consuming and a little difficult. If this could be addressed in an automatic software process after all the initial data is analyzed it would be very helpful.

Response: The calibration has been simplified so that only 3 samples are now needed: PPC, pure rock dust from the mine, and an 80% rock dust mixture (80% mine rock dust, 20% PPC).

f. The scales (to make the initial calibration samples) turn off before you have time to perform the measurements to make your calibration samples. With only one thimble if you lose or crack it, the system is down.

Response: Thank you for informing us about the scales that MSHA ordered and the criticality of having only one cup with the meter. Weigh boats or a piece of paper to hold the dust before placing the components in a sample container for mixing may help provide time before your scale turns off. The commercial meter and kit provides extra cups in case one loses the sample cup.[20]

g. Also it would be beneficial if the meter had a data storage system similar to the new Global Positioning Systems, (i.e. SD card) to keep the calibration for the specific mines once calibrated. This database could be updated as necessary and would save time on recalibrating every time you changed mine locations.

Response: Very good suggestion and one that may be considered as an option in a future version of the meter.

h. Not very sturdy. The switch cover developed a hole the second day of use.

Response: Thank you, the switch and button covers have been changed to a more rugged type for the commercialized version.

---

[20] The commercial CDEM is supplied with large containers for meter calibration. These 4-oz containers provide enough volume to create 30-g or 50-g samples. The small sample cups should not be used for the preparation of a calibration sample.

i. Inconsistent results not in agreement with lab results.

   Response: Thank you for the observations: Since the CDEM and the TIC methods for determining the explosibility of coal and rock dust mixtures use different techniques, it is unrealistic to expect the CDEM to replicate laboratory TIC determinations. Rather, the goal is for the CDEM itself to accurately and quickly identify the potential in-situ explosibility of a dust mixture.

j. The use of the single thimble sized sample cup to scan multiple samples, rock dust, and known mixtures made inspectors worry about contamination of samples. There was always concern the sample cup would be lost. They thought multiple throw away insert sample cups might be better.

   Response: If the sample cup is completely emptied (by tapping on a hard surface), any small residual dust on the walls of the cup, based on lab tests, was not sufficient to indicate a source of error. The CDEM probe is immersed in the dust, picking up reflections from a small amount of dust remaining on the walls of the cup is unlikely. However, the manufacturer is providing a brush to aid in removing the dust from the cup and around the probe and is providing five sample cups in the kit. Therefore, 3 sample cups could be set aside and used with each individual calibration sample only if the user desired.

k. With only one thimble if you lose or crack it, the system is down.

   Response: Multiple sample cups will be supplied and more available for sale.

3. **Do you think that this meter would be useful for rock dust sampling and enforcement (considering that we will have a regulation change to support the Red and Green output)?**

   a. As long as it replaces collecting bag samples.

   Response: Since the CDEM and the TIC methods for determining the explosibility of coal and rock dust mixtures use different techniques, it is unrealistic to expect the CDEM to replicate laboratory TIC determinations. Rather; the goal is for the CDEM itself to accurately and quickly identify the potential in-situ explosibility of a dust mixture. NIOSH researchers believe that the CDEM does a better job of assessing the explosibility of a sample presented to it since it considers the very important factor of the particle size of the coal and rock dust variability whereas the current laboratory ashing method does not.

b. No. Regardless of the outcome an inspector will have to return to an area of the mine that dust samples went out and resample. Right now we only have to observe the area to terminate a citation issued to the area.

Response: Just as your visual inspection of the area from which the sample was collected could not determine inert content without lab analysis, can one determine if adequate rock dust was added to mitigate the potential hazard by visual inspection alone? NIOSH researchers have shown that a visual inspection only for abatement purposes is not adequate to determine if the area has been adequately re-rock dusted. A thin surface coating of rock dust can visually obscure a dangerous un-neutralized sub-layer of coal dust on rock dust or rock dust on coal dust and yet still be a potential dust explosion hazard.

c. Yes, it could prevent fatalities by closing an area until it is properly rock dusted and only terminate when it is safe for miners to return.

Response: Thank you, we agree with your comment. It also underscores the need to be vigilant and up-to-date with rock dusting practices.

d. They would tend to avoid this on larger surveys; however, thought it would be useful to selectively use the meter (possibly even outside the mine) to scan questionable samples and determine if a survey is out. For example scanning 10 bad samples from a 99 sample survey would give them the information they need to cite the survey that day.

Response: Since each sample represents ~ 500 feet of mine entry, 10 out of 99 is perhaps a low hazard weighting factor relevant to the potential consequences if an explosion would occur. Perhaps 1 citation/deficient sample should be considered to encourage interest in maintaining adequate rock dust levels.

e. If used to only analyze questionable areas, this would be a good tool. Did not understand the need of sampling areas that contained up to 6 inches of rock dust, either with the CDEM or by sending samples to Mt. Hope.

Response: Thank you for the very astute observation. Some thought has been given to better focus inspectors' energies on spot sampling those questionable areas as they carry out their routine mine inspection duties.

Although 6 inches deep of rock sounds like a lot of dust on the floor, understanding where the coal dust is distributed in that area is most important for dust explosion hazard assessment. For example, one can have 6 inches of rock dust on the floor with a thin layer (0.01 mm) of fine coal dust lying on top and still have conditions over which an explosion would propagate (Figure E1) [Sapko et al. 1987, NIOSH 2006]. Or there may be an accumulation of fine coal dust on the rib and roof area. If the rib and roof dust represents a nominal entry dust loading of 100 mg/l, there is sufficient dust based on full-scale float dust explosion studies to propagate a coal dust explosion even with 6 inches of rock dust on the floor. Therefore the distribution of the dust within the entry is most important in explosion prevention and hazard assessment. Especially the roof and rib dust.

Figure E1. Cross section of a very thin (0.01-in-thick) explosible float coal dust layer deposited on top of a 3/4-in-thick (20-mm-thick) layer of rock dust [NIOSH 2006].

f. Time consuming because of "double duty." Completing rock dust survey and using CDEM.

Response: Sorry for causing "double duty" but the efforts were most beneficial in answering some of the remaining questions about the CDEM. The double duty work was necessary in this field study for comparing CDEM hazard assessment determinations with results using the conventional explosible assessment approach of sending all samples back to Mt. Hope for analysis.

g. The instrument would be great for companies to check for compliance. If this was used for inspectors to test areas for compliance of areas in question it would be great.

Response: Good point and something that should be considered as part of an effective dust hazard detection and prevention program in using the CDEM.

4. **Any other comments or suggestions for this meter or on our rock dust program in general?**

   a. Very time consuming!

   Response: Completion of this field study did require added burdens and increased the time to complete standard dust surveys and to use the CDEM on all samples. Without the fine efforts of the inspectors, the successful completion of this project would not have been possible.

   b. How is the meter affected by a mine that uses different types of rock dust brands at the same time (bulk and bag)?

   Response: There is a need for tighter rock dust specifications and certification to minimize these types of variations in product quality control issues. That issue is currently being investigated by NIOSH and MSHA. It has been observed that the CDEM adjusts the explosibility assessment to these variations in rock dust particle sizes and, in some cases, is the reason for variance in readings between laboratory LTA analyses and CDEM measurements. In any case, tighter rock dust specification should minimize any differences between bulk and bag product. Until rock dust specifications are tightened and product-certified, the meter needs to be calibrated with the rock dust in use and at least potential variance identified by making up calibrations samples between each source and assessing their impact on meter calibration. Some particle size variations between bulk and bag samples may have little impact on the CDEM operation.

   c. Sometimes the results were spot on and other times they may have been off by 10%-15%.

   Response: This is not surprising since the two methods use different approaches for determining the explosibility of the mixture. Being off when comparing the results of the two tests is one thing. The most important question is: what is the actual explosibility of the sample being evaluated?[21]

---

[21] The differences between the "off" measurements have been extensively discussed in the body of the paper (inadequate mixing of the samples when using the CDEM, moisture remaining in the sample, improper training, calibrating with % rock dust and not % incombustible, etc.). The explosibility of the sample depends on particle size in addition to the incombustible content.

d. Calibration takes a long time.

   Response: Once the calibration samples are created, the calibration of the meter only takes a few minutes. Once the meter is calibrated, the meter should not need to be calibrated frequently. The manufacturer is requiring recalibration of the CDEM after every 200 measurements. It is also necessary to recalibrate the CDEM if there is a change in rock dust being used or you are going to use it at a different mine that uses a different rock dust. Once the calibration samples are made, the calibration of the meter does not take long.

e. Meter needs a case.

   Response: The manufacturer of the commercial meter provides a case with the CDEM.

f. Do we need to carry out our empty vials (the vials and contents may be combustible materials)?

   Response: The vial content may contain a explosible mixture of coal and rock dust just as some of the dust samples bags that you now carry out for subsequent analysis by the MSHA lab at Mt. Hope, WV. The empty unused vials can be used at a later time. In order for the dust in the tube to explode, it has to be dispersed in air and form a flammable dust cloud while in the presence of an ignition source sufficient to ignite the dust cloud. As long as the dust remains in the vial it is not an explosion hazard.

www.ingramcontent.com/pod-product-compliance
Lightning Source LLC
Chambersburg PA
CBHW081857170526
45167CB00007B/3053